Science on the Texas Frontier

D1205937

Dr. Gideon Lincecum on his eightieth birthday.
Courtesy Barker Texas History Center, University of Texas at Austin

Science on the Texas Frontier

Observations of Dr. Gideon Lincecum

Edited by Jerry Bryan Lincecum,
Edward Hake Phillips,
and Peggy A. Redshaw

Introduction by Jerry Bryan Lincecum
Illustrations by Betsy Warren
Foreword by A. C. Greene

Texas A&M University Press
College Station

The paper used in this book meets the minimum requirements
of the American National Standard for Permanence
of Paper for Printed Library Materials, Z39.48-1984.
Binding materials have been chosen for durability.

Library of Congress Cataloging-in-Publication Data

Lincecum, Gideon, 1793–1874.
    Science on the Texas frontier : observations of Dr. Gideon
Lincecum / edited by Jerry Bryan Lincecum, Edward Hake Phillips,
and Peggy A. Redshaw ; introduction by Jerry Bryan Lincecum ;
illustrations by Betsy Warren ; foreword by A.C. Greene. —1st ed.
        p.   cm.
    Includes bibliographical references and index.
    ISBN 0-89096-768-7. — ISBN 0-89096-790-3 (pbk.)
    1. Licecum, Gideon, 1793–1874.   2. Naturalists—Texas—
Biography.   3. Natural history—Texas.   I. Lincecum, Jerry Bryan,
1942–   .   II. Phillips, Edward Hake.   III. Redshaw, Peggy A.
IV. Warren, Betsy.   V. Title.
QH31.L68A3   1997
508'.092—dc21
    [B]   97-10939
                                                                    CIP

In Memory of Sarah Bryan Lincecum
(1797–1867)
&
Sarah Matilda Lincecum Doran
(1833–1919),

Gideon's two "Sallies,"
his wife and youngest daughter,
who shared his passionate love of Nature.

If I cannot have company
whose minds are clearly free,
I would prefer to go alone.
And thus it has turned out with me
through my long sojourn,
I have had no associates,
and my observations and conclusions,
be them right or wrong,
are not trammelled
by the sway of other minds.

—Gideon Lincecum

Contents

Foreword, by A. C. Greene IX

Acknowledgments XIII

Introduction,
 by Jerry Bryan Lincecum 3

CHAPTER 1.
 Gideon and the Ants of Texas 18

CHAPTER 2.
 Texas Botany: Gideon's Catalogue of Useful Plants 51

CHAPTER 3.
 Gideon and the Texas Arthropods 71

CHAPTER 4.
 Creatures of the Air, Land, and Sea 95

CHAPTER 5.
 Texas Geology 110

CHAPTER 6.
 Texas Weather 119

CHAPTER 7.
 From Texas to Mexico 127

CHAPTER 8.
 Gideon in Mexico 140

Appendix 1.
Index of Notable Correspondents   152

Appendix 2.
Edited Version of Gideon's Letters to Darwin   159

Appendix 3.
Thirteen Distinct Species of Texas Grasses   163

Appendix 4.
Gideon's Articles on Geology from *Texas Almanac*   169

Appendix 5.
Published Writings of Gideon Lincecum   179

Notes   181
Selected Bibliography   201
Index   205

Foreword

Texas history likes to honor warriors and politicians; artists, writers, musicians can look elsewhere for recognition—or perhaps find themselves grouped under some all-inclusive heading that covers 150 years and a dozen different forms of their art.

This kind of neglect is doubly true of Texans in science, despite the fact that Texas has a stout scientific history, going back to the sixteenth century when Alvar Nuñez Cabeza de Vaca reported on the flora of South Texas. Samuel Geiser, in his *Naturalists of the Frontier* (1937), lists 152 names of persons who had worked in and contributed to Texas' natural sciences. It is an impressive list that includes French, Swiss, Norwegian, Swedish, British, German, and American naturalists who made Texas their home or spent time in Texas working in their field of scientific interest. Among them were Ferdinand von Roemer, the German geologist; Thomas Drummond, the Scottish botanist; William Kennedy, the British consular officer who wrote the superb two-volume work, *Texas*, containing much natural science; as well as the great naturalist artist, John James Audubon, who, with his son, did work along the Texas Gulf Coast.

Texas was the laboratory for such local scientists as botanists Ferdinand Lindheimer and Charles Wright, and later Julien Reverchon and Jacob Boll, the latter dying in the midst of his Northwest Texas fossil collecting. A substantial number of other collectors and scientific observers lived in Texas, finding and sending new plant, bird, and insect specimens to various European and American universities; but most were looked on with suspicion and often intolerance by their fellow Texans. As Geiser laments, "[They] might have achieved a great deal more if they could have had a more positive stimulus from their environment."[1] This lack of popular recognition for scientific achievement remains today. How many Texans know of Jack Kilday and his scientific importance? Yet he was the developer of the integrated circuit in 1958 which led to the astonishing electronic marvels we use every day.

Perhaps the most interesting of the early Texas scientists was Gideon Lincecum, whose scientific letters and papers form this book. Lincecum, while less schooled than most of the other Texas scientists, was the most colorful and most human, applying his imagination to all sorts of scientific findings. But it is not only his scientific reporting that endears him to the reader; it is his style of writing and self-expression. He frequently wanders off the path while describing his labors, especially in his numerous letters. This irritated some of his more learned correspondents, but it adds savor to our reading today.

In 1967 the late Lois Wood Burkhalter wrote *Gideon Lincecum, 1793–1874: A Biography*, the book which introduced this unusual frontiersman to many readers—including Jerry Bryan Lincecum. I am pleased to say that a review of Ms. Burkhalter's book which I wrote in the *Dallas Times Herald* started Dr. Lincecum on his scholarly research about his previously unknown kin. Jerry Bryan Lincecum, and collaborator Edward Hake Phillips, edited a 1994 volume, *Adventures of a Frontier Naturalist*, using Gideon Lincecum's memoirs, which contain his thoughts and opinions on a great deal more than natural science.

This volume, *Science on the Texas Frontier*, is based on letters and papers of Gideon Lincecum—some letters sent to fellow scientists as famous as Charles Darwin and some papers reprinted in well-known scientific journals. But these are not inscrutable communiqués that speak to us, if at all, in Latin. They are people-to-people letters, letters about birds, insects, and ants. Old Gid loved ants and studied them for years, anthropomorphizing the ants, at times, like a sci-fi writer might humanize alien space invaders. Gideon also loved bees and claimed he had made pets of them, as the reader may note in the opening chapter.

Today, to us, "science" is mostly medical or product and process invention and development; the scientists are in lab coats, not down on their knees watching ants or playing with bees. Some would say that the early Texas naturalists like Gideon Lincecum weren't really scientists, that they were nothing but "bugologists," as some called them. But the dictionary defines science as "knowledge possessed or attained through study or practice" and the scientist as "one learned in science and specially natural science . . . the collector of data through observa-

---

tion and if possible experiment."[2] This fits old Gid like a glove—preferably a deerskin glove that Gideon himself obtained by hunting the deer, tanning the hide, and making the glove.

Gideon's is a decided personality: what he has seen he has seen, what he believes he believes. As he says, ". . . my observations and conclusions, be them right or wrong, are not trammelled by the sway of other minds." But stubborn or not—and he could be both stubborn and biased—he was thought well enough of by his peers. As Samuel Geiser points out, "[Charles] Darwin . . . sponsored the publication of some of Lincecum's researches. . . . The Academy of Natural Sciences of Philadelphia published a paper of Lincecum's filled with data so remarkable that conventional, school-trained naturalists were skeptical of his minute and astonishing observations and unorthodox conclusions."[3]

But Gideon had a weakness which one scientist noted: "The venerable writer had many peculiar notions about society, religion, and the genus homo generally, which he could not refrain from thrusting . . . into the midst of his notes."[4] (Can we really fault Gideon for this assumption of superior information? Look how many scientists today try to tell you what to believe, whom to vote for, or what kind of car to drive.) Regardless of his lack of formal training—he had only five months of enrolled schooling—old Gid lived what he called a scientific life, not only observing and collecting flora and fauna specimens but applying his findings to himself in a unique way: he always slept with his head to the north, kept the windows open winter and summer, went barefoot all summer, and did barefoot mornings, even in the winter.

If Gideon fails to fit the perceived picture of a scientist, he nonetheless benefits from scientific persistence. He was constantly and profitably curious. He wanted to know the "why?" of every unknown circumstance he encountered, whether it be something growing, flying, or slinking away, or the thunder and lightning overhead. He offered his notes and observations freely and in profusion. He was not trained in natural science so much as he was born to it—and in such a manner as will never be possible again, because the natural world will never offer those opportunities again. His scientific value ultimately came not so much from his interpretation of his discoveries as what others were able to do with his work. Even his severest critics freely used his findings.

Gideon did not just represent an area of Texas frontier development, he contributed to a kind of international development. As Geiser says, "The history of scientific exploration of frontier Texas becomes, in a sense, the history of scientific exploration anywhere on a border-line of culture."[5] However, Lincecum and the dozens of other natural scientists involved in Texas study didn't get much governmental support for their work. The only scientists the state supported officially were those trying to find minerals and other resources that would make money.

Although we have not come across a surviving line of his that isn't interesting, we tend to smile at some of Gid's observations and convictions, overlooking the service he and other Texan scientists performed. We must remember, Gideon Lincecum's views of natural science weren't much different from those of the more famous professors and academicians. He wasn't some heretic crying in the wilderness; he was adding to, and helping rearrange, knowledge. As stated earlier, the value of the early seekers and finders comes from the avenues of nature's intentions and perspectives that these scientists opened.

Texas can be proud of them. Closer inspection of their efforts proves to us that we are benefiting, even today, from their crude but powerful conclusions. Their importance lies not in how accurate those conclusions were but in the fact that they were able to make them at all. Texas, their lives declared, is not a province without growth and potential, is not a wasteland, barren of possibilities.

While the soldiers and politicians were expanding our boundaries, the early Texas scientists, such as Gideon Lincecum, were expanding our lives.

—A. C. Greene

Acknowledgments

Many hands have assisted in the preparation of this volume, most especially, of course, the almost tireless hands of Gideon Lincecum, who somehow took the time and the energy, often by candlelight or oil lamp, to pen the remarkable body of letters this volume is but a piece of. He took the pains, too, to make letterpress copies of virtually all of these letters; the surviving letter books, which were donated by family members (especially his youngest daughter, Sarah, and her sons) to the University of Texas, provided the main source for this volume. Our task would have been ever so much more difficult had it not been for the hands of a number of copyists who were provided employment under the New Deal in the 1930s, resulting in the transcribing of each of the letter books. Additional financial support for that project was provided by the Texas Medical Association. Although we found many minor errors in the transcriptions when we compared them with the original letter books, our work in selecting and copying Gideon's scientific letters would have been much more time consuming without the federally financed transcriptions. We owe those anonymous hands and taxpayers a considerable debt.

In making our own copies, we had the ready assistance of the staff at the Center for American History at the University of Texas, headed by Dr. Don E. Carleton, with special courtesy rendered to us by Ralph Elder and his assistants. At Austin College, further copying of selected passages was facilitated by student assistants April Petross, Shannon Straayer, and Suman Reddy. Mary Cole, Sharon Bennett, and Vicki Mansfield, of College Support Services, directed by Cynthia Holloway, were unfailing in their willingness to help. We received considerable support from President Oscar Page and Vice President for Academic Affairs David Jordan, who made funds from the Richardson Endowment available for many of our research expenses. We benefited from generous advice from our colleagues Karl Haller and Light Cummins. Members of the staff of the Abell Library at Austin College, especially Assistant Librarian John R. West, rendered valuable aid in tracking elusive sources and arranging vital interlibrary loans.

We received a windfall of information on Gideon Lincecum's contemporaries in the field of entomology, thanks to the alertness and kindness of scholar W. Conner Sorensen in Wuppertal, Germany (aided by the wonders of e-mail). William Cox, associate archivist at the Smithsonian, provided us with valuable information and materials concerning Lincecum's relationship with that institution. Carol M. Spawn, manuscript/archives librarian at the Academy of Natural Sciences in Philadelphia, responded graciously with documents which show that Lincecum sent the academy many specimens and articles. David A. Pfeiffer at the National Archives and Harold L. Miller at the State Historical Society of Wisconsin also provided valuable information.

Dr. Cheri Wolfe, of Chapel Hill, North Carolina, made our task easier by donating copies of the Lincecum Letter Press typescripts that she had used in her research on Lincecum's Choctaw Manuscript. Damon Bartlett, of San Antonio, a descendant of Gideon's sister Emily Moore, readily responded to our requests for genealogical information. Debbie Lincecum Jensen, of Big Spring, Texas, made us aware that a fossil is named for Gideon. Our longtime friend and one-time colleague Dr. Ron Tyler, director of the Texas State Historical Association, provided much encouragement. His staff's just-published *New Handbook of Texas* arrived in time to ease our task of identifying names and places.

Jerry Bryan Lincecum wishes to recognize his father, Jack Lincecum; his grandfather, Bryan Bowie Lincecum; and his great-grandfather, George Washington Lincecum—all of whom shared and passed on Gideon's love of the natural world. Edward Hake Phillips owes much to his wife, Patricia, for her forbearance and sacrifice during this project. Peggy A. Redshaw wishes to thank her parents, Francis B. (Chick) Redshaw and Margaret Lee Redshaw, for their unfailing support.

—*Jerry Bryan Lincecum*
Edward Hake Phillips
Peggy A. Redshaw

Science on the Texas Frontier

Introduction

BY JERRY BRYAN LINCECUM

On June 19, 1861, a sixty-eight-year-old bearded patriarch sat at his desk in a cabin in the little frontier community of Long Point in Washington County, Texas. He was writing a letter to a fellow naturalist whom he knew only through correspondence, attempting to give a picture of himself and his surroundings.[1]

> I wish I had the capacity to draw a correct portrait of myself; of the full me. . . . I shall not, however, attempt to draw that picture, for I couldn't succeed. But, for better acquaintance, and for your amusement, I may tell you something about my little plans and inventions for employment, as I sluggishly float, like an old sobby log, down the time stream. To enable you to see some portions of the dim picture I am about to scratch the dust from, imagine in the first place that I dwell in a substantial little house composed entirely of red cedar, 40 feet from the main dwelling where the old lady resides; [she] having always, by turns, that gives pleasure to her, a noisy bustle of growing *dears* [grandchildren]; book spoilers and paper scribblers. To get out of the way of these dear little pictures of health and devilment, the above named little house was constructed. In this little house is my bed, books, papers, writing implements, copying press, microscopes, cases of Entomological specimens, Botanical specimens, [as well as] boxes containing various shells, fossils and geological specimens; hanging on the wall are all sorts of hats, non-descripts.
>
> Now go into the neat little latticed gallery, interlaced with running roses, corner posts hollow, full of bees, all acquainted with me and friendly; two or three sleeping on the table now while I write. A large stand [beehive] on the floor, near my bare feet. I don't wear shoes in

warm weather, nor of mornings in winter, until I go round to all my
meteorological instruments, some of which are 50 yards away from
the house, and record the readings of them; then go into the water
closet, bathe the whole surface, scour down with a rough cloth, dress,
and go to breakfast. Eat 3 ozs. bread (corn bread), a spoonful of fresh
butter and a cup of coffee; supper the same; dinner, mostly vegetables
including corn bread, and if any meat, bacon. Sleeping with my head

to the north, windows & doors all open so that every breath taken is a new one. In this manner, I have conducted myself through the last ten years. And during all that time have not enjoyed the luxury of wringing an oosy nose, or suffering any other disease!

But it was of the big bee stand near my bare feet, in the gallery I was speaking. Yes, and there is spread on the floor around it, this warm weather (95 degrees) half a bushel of bees, and I am trying various experiments with them, to ascertain how far I can control these little fellows on the floor, in making them work for me. I first set them to work on one of Sarah's [Gideon's wife] 3 gallon preserve tubs, which she had recently washed and placed to dry a little too near my house—they filled it in 8 days, with honey as clear as spring water. I then cleaned out a little old churn that was lying about the yard, and made them go to work in that. They have been packing into it four days, it will be full in four or five days more. On the opposite side of the stand, I observed another bunch of 100 idle fellows who were spending their time to no purpose. Day before yesterday, I gave them an old tin sardine box—clean—to fill up for me. They seemed to be quite well pleased and are expertly packing it. I am now thinking about giving them a glass tumbler to fill up for me. . . .

At the eastern end of this same gallery, reaching from side to side, is a broad cedar bench, which I use for a lounge. One end of this bench is fitted to the corner post and to the lattice work, as much as 3 feet; and the surplus bees, coming out from the corner post and working round under the bench, which is planked down a little at the sides, are filling the underspace with honeycomb. The roar of all these industrious little insects, going to and from their daily avocations, with the sound of the thousand wings of those whose business it is to ventilate the hives, to a person sitting in the room, is not very dissimilar to the sound of a distant water fall.

So much for the bees. Now raise your eyes, and you find suspended on the walls around, two Hygrometers, one model Barometer, two thermometers, one Ozonometer; and above them again, four old fashioned, Scotch, long lances; relics of ancient warfare, and an assortment of fishing rods. On the floor, lying in great disorder, are various fossils, stone coal, and various paleontological specimens. Look out on the yard; you behold a considerable variety, standing in rows, of the cactus family; some of them as high as your head, with leaves a foot wide or more; and all around them, in rows and in heaps, lie wagon loads of petrified wood; scraps and splinters of a former world in cocoa palms, date palms, cabbage trees, and various other endogens, belonging to the torid zone. They cannot be made to flourish here now. Then raise your eyes, and looking to the south, east and west, you behold the far-reaching prairie, clothed in its rich habiliments of green wavy grass, and flowers, and cows, and corn fields of tall growing corn, scablike harvest-fields, and villas & cottages, which is only terminated by the far off blue curtains of the misty distance.

I have presented you with a badly painted picture of the place and condition, but cannot attempt to portray the man. . . . I am not capacitated to interest society in its present form. I find that my organism is not yet progressed to the plane of polite and fashionable society; and I must content myself with my kindred forms, my brother emmets and my sister worms, a while longer, before I project my head above the surging waves of etiquette and circumlocutory fashions. If I visit, I must stand straight and drill. If I keep away, I can crawl on the ground or climb the trees, studying natural phenomena and developments, from the smallest microscopic animalcule that floats or sails, to the giant forest tree, that waves his dark and defiant limbs in the face of the tornado. There I can perform a small part in the life drama. In elevated society I am a perfect nul[l]ity. But that's nothing.

1. San Augustine
2. Liberty
3. Houston
4. Washington-on-the-Brazos
5. San Felipe
6. Brazoria
7. Galveston
8. Churchill
9. Eagle Lake
10. Goliad
11. San Antonio
12. Bastrop
13. Barton's Bluff (Austin)
14. Long Point
15. Burnam's (La Grange)
16. San Marcos

KEY

GIDEON'S
TEXAS

> There are plenty of that sort [like me], away over behind, in the back
> grounds of the rushing billows of discontented, fine dressed, god serv-
> ing society. I know them.

The writer of these remarkable lines closed the letter as follows: "Fra-
ternally thine, in the grass, Gideon Lincecum." You can find his tombstone in Austin in the Texas State Cemetery in "Founders Row," which is
dominated by a statue of Stephen F. Austin, the "Father of Texas." Lincecum's tombstone announces that he was "A Veteran of the War of
1812," an "Internationally Famous Botanist," and a "Friend of Charles Darwin." It also states that he was "Born in Georgia, April 22, 1793," and "Died
at Long Point, Washington County, Texas, November 28, 1873." There is a
mistake there—he died in 1874—and some exaggeration, but don't
blame Lincecum; the inscription was composed by a committee of his-
torians and carved sixty-two years after his death.

He was indeed a botanist (and much else besides), and he did
correspond with the great Darwin, though not enough to melt the
Englishman's reserve sufficiently to call their relationship that of friends.
The true facts regarding Gideon Lincecum are remarkable enough and
can be found in the 1994 edition of his autobiography, *Adventures of a
Frontier Naturalist.*[2] The reason for a second book of writings by Lincecum
is that in spite of the fact that he was born and raised on the geo-
graphical frontiers of America, in isolation from schools, libraries, and
intellectual colleagues, this man was motivated by his great curiosity
to emulate the philosophes of the Enlightenment and leave behind a
valuable written record of scientific investigation in the form of letters
and papers.[3]

Lincecum's biographer, Lois Wood Burkhalter, wrote:

> Lincecum's early interest in the natural sciences had been developed
> as he trotted along behind his father's ever-moving wagon, darting
> into the forests in pursuit of birds, hunting small game for pets, daw-
> dling in wonder over a new flower, gazing in clear pools at strange
> fish, and crawling on hands and knees to follow the busy traffic of
> small insects. The common names of these wonderful things were so
> long in his vocabulary he never remembered when he learned them;
> his curiosity about such things never grew stale and dull but remained
> as sharp and fresh when he was eighty years of age as in the days of
> his youth.[4]

At the age of fifteen (with only five months of formal schooling), the
young man left home after a dispute with his father and spent three
years working for two different merchants. It was during this time that

the turning point in his education occurred when his employer allowed him to act as the Eatonton, Georgia, sales agent for Parson Mason L. Weems,[5] whose "Flying Library" was a key source of reading material on the frontier. In his "Autobiography" Lincecum wrote of this episode: "My kind-hearted employer permitted me to fit up the lumber-room that was attached to the dry goods store for the purpose. . . . But the greatest thing, in my view of the case, was transforming the lumber room, where I had been sleeping with my borrowed books, into a library of $5,000 worth of the best knowledge that the world could boast. Into this—to me, world of books—I plunged with so much earnestness that I thought or cared for nothing outside of it." When Weems returned to settle accounts, he was impressed with young Gideon's efforts: ". . . besides paying me the full amount of our contract, he presented me with the books I had been studying and some others, which he said suited my years and bent of inclination. . . . Overall, this had been a glorious year for me. It tore open the windows of my abode of darkness, let in the light of science on my awakening faculties and left me with enough of good books on my shelf to nourish and feed the divine flame another year."[6]

Some of his reading material was medical books, which he began studying at the suggestion of Dr. Henry Branham, a friend in Eatonton who had the wisdom to place in his hands Erasmus Darwin's classic work, *Zoonomia, or The Laws of Organic Life*, which had established the elder Darwin as a deep and lasting influence. Lincecum's first letter to Charles Darwin (December 29, 1860) began with an allusion to the elder Darwin's work: "While in my little, quiet office this evening, carefully examining your valuable work 'on the origin of species' &c—[I recalled] the fact that I had received, hot from the press, and read with so much pleasure, the 'Temple of Nature,' which, as far as I know, was the last published work written by your grandfather, Erasmus Darwin." He then goes on to credit the elder Darwin with inspiring him to become a scientific investigator: ". . . it occurred to me, and I was forcibly impressed with the thought that I have been here long enough, all the time in possession of good eyes and ears, long ago set in motion on the plane of investigation, from the suggestions of your worthy Grandfathers inimitable works, to enable me to state some facts relative to your favourite theory of natural selection." Although the term "scientist" was not coined until 1833, Erasmus Darwin was arguably the most knowledgeable scientific investigator and writer of his age. *Zoonomia* was a massive treatise, the first edition in two volumes weighing ten pounds and distilling a lifetime's experience in practical medicine. In his "Autobiography" Lincecum referred to it as "the textbook of practice for the United States,"[7] and it must have been popular, for there were five American editions between 1796 and 1818. Perusal of *The Essential Writings of Erasmus Darwin*[8] made us aware that the elder Darwin's ideas were echoed frequently and significantly in Lincecum's "Autobiography" as well as in his letters and other writings. For example, Lincecum was a teetotaler, and Darwin offered strong arguments against alcoholic drink in *Zoonomia* because of its effect on health. Perhaps the most lasting influence of Erasmus Darwin, and one which ran counter to later nineteenth-century scientific methods, was his willingness to freely engage his imagination for the purpose of writing poetically about science. *The Botanic Garden* and *The Temple of Nature* were effectively compendiums of scientific knowledge in romantic verse, written with a light touch, and they had a profound influence on Lincecum's mode of thinking and writing. His persistence in viewing and depicting the plant and animal worlds imaginatively, poetically, and even humorously on occasion sometimes got him into trouble with academically trained investigators. Several striking examples of his writing in the romantic style of Erasmus Darwin appear in this book (e.g., the description of an ant battle of epic proportions in chapter 1, a mythical account of the genesis of an oak tree that grew in an unusual location in chapter 2, and in chapter 6 his account of traveling aloft in the atmosphere from the Rocky Mountains to Texas in a "Norther-driven car"). We also provide ample evidence that while some of his university-trained correspondents felt that he was out of step with the scientific establishment, nevertheless they read his letters at their meetings, published portions of them, and encouraged him to continue recording observations and collecting specimens for them. For today's readers the result is an interesting paradox: Lincecum's mode of thinking and writing clearly belonged to an earlier, prescientific era, yet he was ahead of his time in a number of areas, such as keeping daily records in minute detail of weather phenomena, experimenting with native grasses to establish conservation practices, and recording the behavior of insects and small animals.

As a case in point, it was Lincecum's imaginative approach to the observation of the Texas harvester ants that gave rise to the "Lincecum myth,"[9] which still appears in the scientific literature. From 1848 until his death (more than a quarter of a century later), he observed these insects, recording in great detail their social life, structure of government, food preferences, sex lives, etc. As an untrained but diligent observer, Lincecum attributed human characteristics to the ants, describing them as having intelligence and courage, and his reports were doubted by the "high lads," as he called the academically-trained scientists. In particular, his contention that the ants plant and harvest their favorite grain was labeled the "Lincecum myth" and set off a bitter controversy among eminent naturalists. In a similar vein, when members of the Philadelphia Academy responded with disbelief to Lincecum's account of an epic ant battle, his response to Elias Durand (June 16, 1866) was caustic:

> . . . we must not entertain any feeling on account of the unbelief of these men, as they perhaps have seen only the dry specimens of the ants in question, and therefore could not know anything in relation to their living character and belligerent habits. These cabinet investigators have seldom felt the tick bites and stings of the warrior ants in the forests and plains traversed by the industrious and earnest examiner of nature. Let them try the experiment of going out and looking for themselves, and if they seek earnestly, they will find many things true which their present mental condition would stile poetical and imaginative. They would find that ants do have wars, and that twenty or forty thousand of some species of them, engaged on a single battlefield, is nothing for the initiated to wonder at.

The controversy continued long after his death. In Bert Hölldobler and Edward O. Wilson's *The Ants* (1990), Lincecum's "famous misconception" about the harvester ants is highlighted, and they report that an intense revival of interest in these ants began in the 1970s, resulting in both field and experimental studies on foraging and competition that have "grown into an important chapter of modern general ecology."[10] The fact that his ants are still being studied and that his observations are regularly cited in the literature would please Lincecum immensely.

The period during which Lincecum was most active in scientific investigation was, according to one historian of science, a formative period: "In general the mid-nineteenth century has been a neglected period in American science. . . . Throughout the century, American scientists in most fields were typically regarded at best as junior partners of their European betters. If American contributions were recognized at all, it was not until the closing decades. By the 1870s, however, . . . America had become a 'mature but small scientific nation with many of its basic institutions in existence or in embryo'; its scientific infrastructure . . . had been formed and 'launched' in the three decades preceding."[11] It was in the 1860s and early 1870s that Lincecum was fired with great enthusiasm and worked most diligently. In more than one entry in his diary for 1867 he berated himself for having done little or nothing that day to contribute toward the advancement of the natural sciences.

In an unpublished paper presented before the East Texas Historical Society in 1994, Peggy Redshaw posed the question, "What did science mean to Gideon Lincecum?" In answering this question, Redshaw drew attention to three of Gideon's pronouncements:

- *"How little people see of the things they are daily trampling over."*
- *"Some people think and some have already said it—that I am deranged on the subject of bugs, ants, frogs, etc. Well, it is at least a harmless derangement."*
- *"If I can not have company whose minds are clearly free, I would prefer to go alone. And thus it has turned out with me through my long sojourn, I have had no associates, and my observations and conclusions, be them right or wrong, are not trammelled by the sway of other minds."*

Redshaw stated:

As a true child of the Enlightenment, Gideon found truth and logic in science. He viewed science as a profession higher than religion or politics, never doubting it would prevail over the traps of ignorance and superstition and would perfect the human race. His pursuit of the ants, bees, beetles, flowers, etc. was an invention for his mental and physical enjoyment and personal happiness. Gideon recorded almost everything he observed and pushed towards conclusions— which he called 'ultimates.' His curiosity never grew stale and his interests stretched from birds, animals, fossils, fish, insects, grasses, geologic formations, mineral deposits, meteorology, to name a few of his many, many endeavors. He had a place to keep collections of the shells, fossils, geologic specimens, plants, petrified logs, bones, jars of insects—the red cedar house, which served as a combination office and laboratory, where he could retreat from his large family. It was his private space. There he had his letter press, and quiet time to record his observations about daily life in the natural world.

There he penned many hundreds of letters to his scientific correspondents as well as a lesser number to family members, friends who

were not scientists, and those who wrote him about medical treatment or some miscellaneous topic. Consistent with the letter-writing habits of nineteenth-century people generally, Lincecum wrote letters that were often quite lengthy, some of them running on for ten pages or more. For an indication of how important his scientific letter writing was to him, consider this comment he wrote in 1862 regarding the cessation of it during the Civil War: "The correspondence which had so long subsisted with scientific friends north of the Masons and Dixon's line had been perminantly interrupted, and I was at a loss to know where I could obtain even a moderate supply of proper aliment [nourishment] for my old starving mentality."[12] In fact Lincecum was staunchly proslavery and so antiyankee that in 1860 he had doubts about giving specimens to Northern scientists, according to a letter addressed to S. B. Buckley (December 19, 1860): "If those fellows in Philadelphia [members of the Academy of Natural Sciences] should turn out to be of the Black stripe, what good reason is there why Sarah [his daughter] and myself should not bestow our 800 specimens on our own state?" Not only did the Civil War temporarily disrupt his lines of communication with Northern scientists and shake his confidence in them; it also resulted in the destruction of a letterpress volume containing copies of 900 pages of correspondence. Most importantly, the outcome of the war left the septuagenarian penniless, forcing him to turn aside from science and give his foremost attention to making a living.[13] He wrote to Buckley (January, 1866, undated): "When the war broke up, and I had time to look around, I found I had lost everything but the homestead. It is the truth, that I did not have a blanket, nor horse, no mule nor jackass, left to my name, nor one single dollar." But he did manage to resume the correspondence with his fellow scientists, and he overcame his scruples about sharing his knowledge and specimens with Yankees. Burkhalter wrote: "Gideon had long advocated a state collection and scientific academy in Texas. There being none, he was anxious that his work be preserved; so he presented his collection of botany from throughout Texas to the Philadelphia Academy."[14] After the war Northern scientists deluged him with requests for collections and information, and he donated extensive collections to the Smithsonian,[15] as well as sending numerous specimens to individuals.

In addition to Charles Darwin, the noted scientists with whom Lincecum corresponded included the physicist Joseph Henry, who initiated the first United States weather reporting system; the botanist and geologist Samuel Botsford Buckley, who served as the state geologist of Texas for a number of years before and after the Civil War; the botanist Elias Durand, who was a member of the Philadelphia Academy of Sciences; the ornithologist Spencer F. Baird, who joined the

Smithsonian in 1850 and as director of the National Museum worked to build its collections; the entomologist Horatio Charles Wood, a physician, naturalist, and professor in the medical school of the University of Pennsylvania; the paleontologist Joseph Leidy, chairman of the board of curators of the Academy of Natural Sciences; the entomologist Alpheus S. Packard, Jr., president of the Essex Institute of Salem, Massachusetts, and publisher of the *American Naturalist* (which printed edited versions of a number of Lincecum's letters); the paleontologist Edward Drinker Cope, who visited Texas after Lincecum's death and published a number of papers on Texas vertebrates; the entomologist Ezra Townsend Cresson, publisher of the *Practical Entomologist*, who had lived in New Braunfels, Texas, for a time and worked with the Texas botanist Lindheimer; and the meteorologist and botanist George Engelmann, who was interested in Texas grapes. Each of these is identified in more detail in appendix 1, Index of Notable Correspondents.

In exchange for masses of information and hundreds of specimens, all Lincecum expected was respect, courtesy, and, occasionally, some words of instruction or a helpful book or tool. Spencer Baird and Joseph Henry supplied him with a number of books and pamphlets, as well as some supplies for use in collecting birds' eggs, and after the Civil War Lincecum's old friend Elias Durand sent him Agassiz's *Introduction to the Study of Natural History*, Packard's *How to Collect and Observe Insects*, Wood's *Myriapoda of North America*, Chapman's *Flora of the Southern United States*, and Gray's *Botanical Textbook*.[16] Unfortunately, not everyone reciprocated Gideon's generosity, and his worst treatment was suffered at the hands of Horatio Charles Wood. After the Civil War, Wood asked Lincecum to collect insects of all kinds, and Lincecum agreed, suggesting that with the aid of three small boys he could supply an extensive collection. Having come late to the study of insects, he wrote Wood (July 3, 1866):

> I must have a work on entomology, or I can't get along very well with my bugs. Buckley wrote me the other day, that Harris' treatise on insects, 'On Entomology' with plates, 2nd or last edition, Edited by Dr. Fitch, is the best work. Please send me a copy of that work, or any other *better* work. . . . I see not why I may not, during my leisure hours, look into and learn the names of the great bug family, before I go. —I know them all, or nearly so, with the history of their mode of life, manner of procuring food, procreative processes, and how, many of them, contrive to have a great deal of recreation and amusement, and all this without having names for them. I do, to be sure, have names for the different orders and classes, but not many names for species. Yet I know them all apart and what the individuals do to get a living. I

suppose I might have been able to repeat the scientific names of the individuals if I had possessed the books belonging to the subject. But I did not, nor have I ever seen a work on Entomology, to this day.

After Wood received Lincecum's collection numbering several hundreds of insects and did not answer, Gideon wrote again (December 6, 1866): "Books would hinder me now. I shall continue to discribe nature as I see it. What I may so write, will, of course, correspond with no book and therefore can be of no use to the book men. But it will remain as a curiosity and someday some *Man* will see it. . . . Cooped up as you are in a large city, with time occupied professionally, it is almost impossible for you to be anything of a field naturalist." Wood failed repeatedly to acknowledge the old man's requests for the thirty dollars he had promised to pay the three orphan boys who served as his helpers.[17] After nine months Gideon gave up and paid the boys himself, although he could ill afford it, remarking of Wood: "He is a small man."[18] When Lincecum read in the *American Naturalist* that Wood had presented a paper entitled "Description of a New Species of Texas Myriapoda," he had to admit that Wood was "an industrious student" and then commented in a letter to Durand (October 16, 1866):

> But the labor was greatly curtailed by his finding his Myriapoda and Pedipalpa already collected and arranged to hand in the Smithsonian. If he had to go out and hunt them in the deep forests, tangled vines, brushy gullies, snaky brooks and grassy fields, among the ticks and red bugs, and then have to roll old decaying logs, and bark the dead trees to collect specimens of his favorite reptiles, it would consume a little more time and manhood. At this moment, I feel the itchy gnawing of perhaps 100 redbugs on my surface. They are not so large as a pen dot and their bite to some people is very painful. But the rich remuneration is found in seeing the home of the live specimens and his mode of keeping house.

For an indication of Lincecum's generosity and willingness to share his findings, consider his reply (July, 1866, undated) to a query from Wood about quoting from his letters in publications: "As to the privilege of publishing extracts from any of my letters, if you find anything that will benefit science, or strengthen the spirit of investigation I dare not object to their being published. Under this head, it is perhaps no more than proper for me to say to you that whatever I may write, as a statement of any facts in natural history, may be relied upon as the unbiased results of a serious and solitary investigation, with no other aim or object but to arrive at truth." In response to his many valuable contributions resulting from this spirit and attitude, the members of the Philadelphia Academy of Natural Sciences voted unanimously on March 26, 1867, to elect Lincecum a corresponding member.[19] This was the highest honor and recognition that they could give to an amateur, and it was not passed out indiscriminately.

Furthermore, in March, 1874, when Gideon was eighty years old, he received a letter from the prominent naturalist, Elliot Coues of Washington, D. C., that must have brought him great satisfaction. Coues wrote:

> Although personally a stranger to you, I am not so to your numerous interesting writings on the habits of animals, and the freedom of intercourse which all good naturalists encourage among each other will I doubt not be extended to me by you on this occasion. . . .
>
> My chief deficiency is in the matter of accurate and minute information respecting the habits of the smaller animals—the rats, mice, gophers, squirrels, shrews, moles, &c. &c. Having seen all your publications on these subjects, in the *Naturalist &* elsewhere, I am convinced that there is no one in the country who has paid more attention to these things than yourself, or who has studied more successfully. And I am anxious to avail myself of your investigations.
>
> I have lately completed and am about to publish an elaborate monograph of the *Muridae*, and you need hardly be told how very valuable were the numerous specimens which I found in the Smithsonian from you.[20]

There is a notation on the Coues letter stating that Lincecum answered it on May 4, 1874 (only six months before his death), but the reply is not found in the Lincecum papers.[21] Although the infirmities of old age prevented him from contributing directly to Coues's work, there is ample evidence in the 1877 Hayden Survey collection of monographs (*Report of the United States Geological Survey of the Territories*, XI), written by Coues and Joel Asaph Allen, that his earlier publications and the collections he had made for the Smithsonian were valuable. In the bibliographies as well as the text of this volume there are several citations of Gideon's published articles, and he is listed as the donor of a number of specimens that Coues and Allen examined as part of their research.[22] Compared with the output of naturalists like Joseph Leidy, Edward Drinker Cope, and Spencer Baird, Lincecum's contributions are modest, but considering his isolation and the lack of any formal training it is surprising that he was able to contribute at all and astonishing that his observations earned the high regard of someone like Coues.

An objective assessment of Lincecum's contribution to one field of science, that of entomology, was offered in 1995 by W. Conner Sorensen in *Brethren of the Net: American Entomology, 1840–1880*. Using A. S. Packard's annual *Record of American Entomology* for the years 1868–73 to make a detailed analysis of "the leadership of the American entomological community" of 1870, Sorensen ranked Lincecum as fiftieth in priority (or overall importance for American entomology) among 108 Americans and Canadians cited. Surprisingly, this placed the Texan higher than a number of his fellow scientists with whom he corresponded: Joseph Leidy (ranked fifty-seventh), George W. Peck (fifty-eighth), and E. D. Cope (sixty-fourth).[23] On January 13, 1861, Lincecum wrote Durand, ". . . anything I can do in my small way that will add a single line to the record of the natural—the only true sciences—shall be done as free as the water flows." The publication of a selection of his scientific letters and papers in the present volume attests that he deserves credit for adding more than a few lines to the record.

Chapter 1

Gideon and the Ants of Texas

Go to the ant, thou sluggard; consider her ways, and be wise:
which having no guide, overseer, or ruler, provideth her meat
in the summer, and gathereth her food in the harvest.

Proverbs 6:6–8

Shortly after he relocated his family in Texas in April, 1848, Gideon Lincecum began to give serious study to the local ant population. By the time he read Charles Darwin's *On the Origin of Species*, in December, 1860, he had been observing Texas ants for twelve years and writing out his findings, and in one species in particular he saw evidence to support the theory of natural selection. Here is the full text of his first letter to Darwin, dated December 29, 1860:

> Charles Darwin M.A.
> Down, Bromley. Kent. England.
>
> Dear Sir.
>   While in my little, quiet office this evening, carefully examining your valuable work "on the origin of species" &c— [I recalled] the fact that I had received, hot from the press, and read with so much pleasure, the "Temple of Nature," which, as far as I know, was the last published work written by your grandfather, Erasmus Darwin. And now, with emotions of approving interest, [I] find myself profitably perusing the highly useful labours of his truly investigating Grandson,

who, himself alludes favorable to having a grown up son; it occurred to me, and I was forcibly impressed with the thought, that I have been here long enough, all the time in possession of good eyes and ears, long ago set in motion on the plane of investigation, from the suggestions of your worthy Grandfathers inimitable works, to enable me to state some facts relative to your favourite theory of natural selection that may not be offensive to you. With that thought by natural selection . . . came the desire to communicate,—not much, for writing has become irksom to me—a few facts of striking character, in natural movements, which are, as I conceive, in harmony with your theory. You record many cases of equal force; mine will be a few more, and at the worst will not damage your labors, as you can, very conveniently, avoid their consideration.

In my journal of observations I find many cases applicable to your theory of natural selection, but in my present state of mind, I feel more inclined to state some of my observations on the agricultural successes of one of our many species of Texas ants. It may interest you some. It cannot injure. Consider it. If you like it, and want more, write.

The species of Formica,[1] which I have named Agricultural, is a large brownish-red ant, dwells in paved cities, is a farmer, thrifty and healthy; is dilligent and thoughtful, making suitable and timely arrangements for the changing seasons; in short, he is endowed with capacities sufficient to contend with much skill and ingenuity, and untiring patience, with the varying exigencies which he may encounter in the life conflict.

When he selects a situation, upon which to locate a city, if it is on ordinarily dry land, he bores a hole, around which he elevates the surface three, sometimes six inches, forming a low, circular mound,

with a very gentle inclination from the center to its outer
limits, which on an average is three to four feet from the
entrance. But if the location is made on a low, or flat wet
land, liable to inundation, though the ground may be perfectly dry at the time he does the work, he nevertheless
elevates his mound in the form of a pretty sharp cone, to
the hight of fifteen to twenty inches, sometimes even more,
having the entrance near the apex. Around this, and its the
same case with the upland cities, he clears the ground of all
obstructions, levels and smoothes the surface to the distance of three or four feet from the gate of the city, giving it
the appearance of a handsom pavement, as it really is. Upon
this pavement not a spire of any green thing is permitted to
grow, except a single species of grain bearing grass. Having
planted it in a circle around, and two or three feet from
the center of the mound, he nurses and cultivates it with
constant care, cutting away all other grasses and weeds that
may spring up amongst it, and all round outside of the farm
circle to the extent of one or two feet. The cultivated grass
grows luxuriently, producing a heavy crop of small, white,
flinty seeds, which under the microscope very much resemble the rice of commerce. When it gets ripe it is carefully
harvested, and carried by the workers, chaff and all, into the

grainery cells, where it is divested of the chaff and packed away; the chaff is taken out and thrown beyond the limits of the pavement.

During protracted spells of wet weather, it sometimes happens that their provision stores become damp and liable, as they are invariably seeds of some kind, to sprout and spoil. If this has occurred, the first fair day after the rain they bring out the damp and damaged stores, expose them to the sun til they are dry, when they carry back and pack away all the sound seeds, leaving all that are sprouted to waste.

In a peach orchard, not far from my residence, is a considerable elevation, on the top of which there is an extensive bed of rock. In the sand beds overlying the portions of this rock are five cities of the agricultural ants. They are evidently quite ancient cities and may have occupied this elevated rock for thousands of years. My term of observations on their manners and customs has been limited to the time the cattle, by the orchard enclosure, have been kept away from their Rice farms 12 years. Those cities which are outside of the inclosures, as well as the protected cities, are, at the proper season, invariably planted with ant Rice.[2] And we accordingly see it sprin[g]ling up in the farm circle about the first of November, every year. Of late years however, the cows are eating out the grass much closer than formerly, preventing the ant farms and everywhere else from maturing seeds. I notice that the agricultural ants—but no other species of ant—are locating their cities along the turn rows in the fields, walks in the gardens, inside about the gates &c where they can cultivate their farms without molestation from the cattle. And here, if it was not for the dread I feel of offending that host of investigators, who know all things and who are satisfied there is no truth beyond the range of their own observations, I should be almost wicked enough to state it as my opinion, that these granivorous agriculturalists had intentionally gone in and unobtrusively settled themselves in the unplowed portions of our inclosures for the purpose of escaping from the depredations of the cattle. This species of ant do no damage to the farms, however numerous.

There can be no doubt of the fact, that the peculiar grain bearing grass, mention[ed] above, is intentionally planted; in farmer-like manner, the ground upon which it stands, is carefully divested of all other grasses and weeds during the time of its growth, and that when it is ripe and the grain taken care of, they cut away the dry stubble and, carrying it out of the way, leave the pavement unincumbered until the insuing autumn, when the same ant rice, and in the same circle appears again, receiving the same agricultural attention as did the previous crop, and so on, year after year, as *I know* to be the case, in all situations where they are protected from granivorous animals.

I have no theological clique to curb the play of my mind, such as it is, nor is it fettered to the cosmogony of any priestly documents, ancient or modern, and so, I may here suggest that it may not exceed a few short millions of years, since this now, bold, healthy, thrifty, city paving, agricultural species of emmet belonged perhaps to a feeble variety of hunters and herders who, in the struggle for existance, were entirely dependent on the natural droppings of the seeds of vegetation, such insects and worms as their hunters could master, and the Aphi[d]s for subsistence. Then the "struggle for existence" was difficult, alike to all the communities of that variety. At a certain period, however, and near the gates of a particular city, a seed of the favourite grain bearing grass by carelessness had been left—or if you like, and it does not overreach the bounds of their intelligence in my estimation, some thoughtful ant planted it there for experiment. Be this as it may, the seed sprouted and grew up to a flourishing tuft, producing a bounteful crop of the favourite Rice, which in due time was, by the inhabitants of the city, carefully harvested. This, in addition to what had been collected from the country around, smartly enlarged their stores of food, whereby greater numbers would survive in the winter; in greater strength they would be able to compete in the *struggle for life* during the coming season, again a few plants of the rice would be tried, the results, of course, would be encouraging, and thus by little and little the agricultural habit gradually obtained. By the good effects of less labour and more food, the population of the city were steadily increasing in form, size, vigor, and numbers. And so, the slow but sure modifying principles and powers of natural selection, rolled on, gathering as it progressed through an unimaginable series of ages until they found themselves occupying a great paved city, surrounded by a stately farm and crowded with a lusty, warlike population, colony after colony had gone out, and carrying with them the agricultural habit, they had dotted the far off regions with their robust, dominant farm cities; all so changed in strength, habits, and general appearance, that they bore no similarity to the type of the original variety from which they had branched. But there were none of the parent variety to be found for comparrison.

They were extinct. Had been exterminated by the forces, under favorable conditions, of natural selection. The no-corn-making, pastoral *Abel* had been overcome—rooted out by the vigorous, prolific, agricultural *Cain*. Thus the transitional link between this now distinct and wonderfully developed species, and probably the *well* digging Species, who are about the same size and colour, subsisting on the foliage of vegetation, are to Science forever lost.

I might ennumerat twenty odd Species of Texas ants. The two species I have alluded to are, however, the most interesting. But I must abstain from any attempt at giving you even a glimpse of the history of their stupendeous public works, governmental systems, laws, civil and military, slaves, prisoners, distructive wars &c &c. It would require a year or two, to perform the work in the stile it merits.

Old men, like myself, can do but little of anything, and ought not to promise themselves much, but I think I will, sometime when I feel in proper condition for it, write out a short account of what I have seen during the past twelve years amongst our abundant animal and vegetable fossils, our insect Kingdom and our blooming prairies.

I am a native American, born in the smoke of our first revolution; was raised and have always been a dweller in the wild border countries, and now, I think the chances are pretty good for me to make my exit in the turmoil and smoke of another revolution [i.e., the Civil War].

Most Respectfully,
Gideon Lincecum

Within two months he had received a reply, and although the letter from Darwin has not survived, Gideon's response (dated March 4, 1861) gives a good indication of its contents.

Your kind letter of 27th January, was just one month on the way. I would not pester you again, but for the question contained in it. You speak dubiously of my "long career in wild countries." I might do the same in regard to your opper-tunities in the *tame* country of books and seminaries, pomp-ous priests and legal superstition. But I don't, except your trip around the globe. If I can not have company whose minds are clearly free, I would prefer to go alone. And thus it has turned out with me through my long sojourn. I have had no associates, and my observations and conclusions, be them right or wrong, are not trammeled by the sway of other minds. Except five month's schooling at a deserted log cabin in the backwoods of Georgia by an old drunkard, my mind has not been biased by training of any kind from designing man. In the cane brakes and unhacked forests on the borders of the above named state, with the muscogee Indian boys for my classmates, I learned my first lessons in nature's grand seminary. Here arose my first thoughts on the subject that is now, by yourself, and by me subscribed to, denominated "Natural Selection." Pardon this digression.

You ask, "Do you suppose that the ants plant seeds for the ensuing crop?"—I have not the slightest doubt of it. And my conclusions have not been arrived at from hasty, or careless observation or from seeing the ants do something that looked a little like it, and then guess at the results. I have, at all seasons, watched the same ant cities, during the last 12 years, and I know what I stated in my former letter is true. I visited the same cities yesterday. I find their crop of *ant Rice* growing finely; exhibiting also the signs of high cultiva-tion; not a blade of any other kind of grass or weed, to be seen in 12 inches of the circular row of the Rice.

We have not only agricultural ants in Texas, but we also have a species, that are regular Horticulturists. The mound, which is constituted of the sand that is thrown out from their cells below and their extensive tunnels, is from one to two feet high, and sometimes spreding over an area of two or three square rods. It is upon this elevated ground that they plant their shade trees. They cannot stand our summer sun, and in those cities, of recent date, not having had time to grow the shade, no ant is ever seen on the mound, when the sun is 9 o'clock high in hot weather. They are compelled to perform their work during the night time, until the city is properly shaded. Neither can they travel out over the un-shaded plains of a sunny day, to bring in the provisions for the labor workers, hence the necessity of tunnels or under-ground passages to the trees and patches of herbaceous plants that produces the leaves of which the food of this species entirely consists. The sand which is taken out from their tunnels is all thrown out upon the city mound. The excavation of these underground passages are always commenced in the city and extending outwards to some district—often four or five hundred yards—that produces plentiful crops of the kind of leaves upon which they feed, the amount of sand thrown out from them increases the elevation of the mound very considerably; the bore of the tunnel, for the purpose of allowing sufficient room for them to carry a piece of leaf through it, that is as wide as a dime, sometimes larger, is generally an inch in diameter. Its outer terminus, most commonly, ends at two or three points, under the shade of a spreading tree, or in a garden or a cornfield. The holes where they come out are always con-cealed, by being carefully and ingeneously covered with dry leaves, bits of stick &c. When they enter a garden in this way, they seldom fail to ruin it, in spite of the efforts of the owner to prevent them. All kinds of fruit trees, many flowering shrubs and garden vegetables are entirely trimmed of their leaves and totally ruined. One gentleman on San Antonio River had a very fine garden invaded by the cutting ant (the common name of the species in question) in great numbers, which were rapidly destroying his vegetables by night. As he had good irrigating facilities he conceived the idea of surrounding his garden by a considerable ditch and let a sluice of running water through it. All of which he accomplished in good stile, and for two or three days was boasting that he had ou[s]ted the little pests. It was not long til he found that his garden was being cut up and damaged as bad as before he had surrounded it with water; but as they perpetrated their mischief at night, he could not discover how they had managed to cross his ditch. It was several days before he found the secret and then by accident found, at the root of a little flowering shrub in the garden, the concealed terminus of one of their tunnels. On further search he discovered several other concealed holes and now form[ed] a resolution that he would not be outdone by them, but that he would destroy them, cost what it might. So calling four or five negroes with their implements for ditching, set them to work, ditching and following the ant tunnels from inside the garden; he found that the tunnel passed at a dry depth beneath the water in the ditch, when it rose again within eighteen inches to two feet of the surface. At about that depth, he followed it to a large mound city of ants, distant four hundred yards. And now the only chance to destroy them was to dig out the city. This he was prosecuting at last accounts.

Their horticultural action is exhibitted in their nice judgement in the selection of the quick growing species of the heavy foilaged trees to plant for shade, of which they cultivate four or five kinds *Celtis occidentalis, Viburnum dentatum, Ilex vomatoria, Zanthoxylum carolinanum*, and the mustang grape vine;[5] all beautifully ornamental. When they locate a city so far out on the bald prairie that they cannot carry the seeds of the above named trees so great a distance through the grass, they collect and plant the seeds of the *Argemone mexicana*,[4] a large, quick growing prairy weed, having spreading tops with large leaves, which under the ant culture, besides producing ample shade, is, with its large white flowers, quite ornamental.

It would require a considerable volume to discribe this most interesting type of our ants. I will not bore you further on that topic. I have answered your single question; you must excuse me.

You may howe[ve]r, say to your brother, that if he feels like it, and will put the proper questions to me, he will find, that these little Emmets can teach lessons to the genus homo even, that would be profitable to imitate.

You may be too much engaged, but if you feel any interest in it, this correspondence need not cease.

I Remain Dear sir
Very truly thine
Gideon Lincecum

The correspondence did cease, perhaps because of the Civil War, but on April 18, 1861, Darwin read portions of the two letters to the Linnean Society, and an edited version of them was later published in the journal of the society. *A Calendar of the Correspondence of Charles Darwin*[5] enabled us to discover that first Darwin forwarded the two letters, with annotation and a cover letter, to George Busk, undersecretary of the Linnean Society, asking for help in deciding whether to read them before the society. From the Darwin Letters Project at Cambridge University we obtained copies of Darwin's cover letter and the annotated versions of Gideon's two letters that he transmitted to Busk. Here is the text of Darwin's cover letter, as transcribed from a xerox copy of the holograph original:[6]

Letter to Accompany Lincecum, Agric. Ant of Texas
Ap. 5   Down, Bromley, Kent

Dear Busk

In last no. of Journal of Linn. Soc. there is a marvellous account of ants; enclosed is its match, which if you think fit, might be read, with some such title as "Extracts from two Letters from G. Lincecum, Esq., of Long Point, Texas, to Chas. Darwin, Esq., on the habits of ants."—Please observe, I know nothing of writer.—But if you will take trouble to read the *whole* of these extraordinary epistles, I think you will be impressed into belief that the man does not *intentionally* tell lies. He paid the heavy postage on both.—I have struck out with pencil what ought not to be read.—If the facts are true, it is perhaps most marvellous instinct ever recorded.— Really I can almost believe the statements, after Kirby's account of the ants bringing up the eggs of their imprisoned aphids to the sun to be warmed and to be hatched early that they might be milked soon.—
The only use of publishing such a paper in my estimation is that it might call some other observer's attention to these points.—
The whole letters are so odd that they are almost worth your reading,—such spelling—such grammar!
He evidently speaks the truth that he was never educated.— You must use your own judgment whether to read it,—I hardly know what to think.
Plz relieve me from my uncertainty.

                                                                Darwin

P.S. In reading the 1' [primary] letter, attend to his paging; for the order goes very oddly.

Evidently someone relieved Darwin of his uncertainty, for in 1862 an edited version of excerpts from Gideon's two letters appeared in the Linnean Society's journal, under the heading: "Notice on the Habits of the 'Agricultural Ant' of Texas." The credit line read: "by Gideon Lincecum, Esq., M.D. Communicated by Charles Darwin, Esq., F.R.S.,

F.L.S. Read April 18, 1861." The article began with a note: "The following is merely an abstract of Dr. Lincecum's communication, containing only what appears to be most remarkable and novel in it in the way of observation." Since the article is repetitive of the two letters reproduced above, we have chosen to omit it here and make it available in appendix 2 for those who wish to compare it with Gideon's letters. As to why Darwin found Gideon's letters so interesting and chose to sponsor their publication, W. Conner Sorensen has some insight: "In the years immediately following the publication of the *Origin*, Darwin paid increasing attention to insects. He corresponded extensively with entomologists, asking questions, suggesting solutions, and encouraging them to pursue evolutionary themes in their own observations. In the 1860s he used the meetings of the Entomological Society of London as a sounding board for his developing ideas on natural selection. The result of Darwin's intense interest in entomology in this period is apparent in *The Descent of Man* (1871), which draws upon entomological sources for much of its supporting data."[7] Just as Lincecum suspected, his accounts of the ant behavior offered Darwin supporting evidence for his evolutionary theory.

Several months before he first wrote to Darwin, Lincecum began a long and intimate correspondence with Elias Durand, a botanist and member of the Philadelphia Academy of Sciences, and eventually it too resulted in the publication of notable papers on Texas ants. The two men were close in age, as Durand was born in France in 1794. Commissioned a pharmacist in the French army in 1813, he participated in numerous battles of the Napoleonic Wars. He left France in 1816 because of his Napoleonic sympathies and in 1825 opened in Philadelphia a pharmacy which became an informal clubhouse for scientists. He was elected in 1824 to the Philadelphia Academy and to the College of Pharmacy, and in 1854 to the American Philosophical Society. When his second wife died in 1851 he turned his pharmacy over to his son and devoted the remainder of his life to botanical studies. In 1868 he gave his collection of over 100,000 botanical specimens (including many obtained from Lincecum) to the *Jardin des Plantes*, Paris. Durand died August 14, 1873, just one year before Lincecum's death.[8]

After exchanging several letters with Durand, Lincecum appraised him as follows in a letter to S. B. Buckley, dated June 10, 1860: "From his autograph, and the style and manner of his address, I am led to the conclusion that he is a gentleman of high-pitched type; a type of rear [rare] occurrence, and as such, I have enrolled his name amongst my favorites." In a letter to Durand dated March 3, 1860, Lincecum offered a sketch of his background:

As I am quite and [sic] old man, you must not expect too much from me. Being 70 years of age [actually he was 67], I am somewhat clumsey, both physically and mentally. Added to all this, I have always dwelt in the border countries, and what little I know, in the way of book learning, is quite limitted. I have spent the greater portion of my long life in the forests with the singing birds, blushing flowers, rocky cliffs and clear rippling streams, in company with the unsofisticated hunter and truth-speaking Indian tribes of the Southern frunteers. I am, however, healthy and always sober, so that whatever I may do in the way of contribution to your institution or to science anywhere shall be reliable.

Lincecum recognized that his lack of training and isolation from other scientists placed him at a disadvantage. He wrote Durand on February 6, 1861: "Having had no scientific associates, either in men or books, worth naming, I am not posted as to how much the world knows of Natures great record. They may know it all; and that that looks so fresh and beautiful to my untutored mind, and which I prise so dearly may all be very commonplace to the scientific investigator." One of his more poignant comments on his isolation occurred in a letter to Durand, dated February 22, 1866: "I have no associates about me to encourage me or keep alive the spirit of inquiry at all. Adam, in the year one, Eden, was no more alone, in that respect, than I am."

At Durand's request Lincecum, aided by his youngest daughter, Sarah,[9] devoted considerable time in 1860 to gathering a sizeable collection of Texas botanical specimens for the Philadelphia Academy. In the academy's *Proceedings* for May, 1861, appears this item, dated May 7: "The report of Mr. E. Durand upon the collection of plants recently received from Dr. Lincecum of Long Point, Texas, in which he congratulates the Academy upon having acquired in Dr. Lincecum a zealous and useful correspondent, was read."[10]

Lincecum's letters make it clear that because of growing war clouds and his suspicions that Northern scientists might be supporters of "Black Republicanism," he had doubts about sending the collection to the Academy. He wrote his fellow Texan Buckley (December 19, 1860): "If those fellows in Philadelphia should turn out to be of the Black stripe, what good reason is there why Sarah and myself should not bestow our 800 specimens on our own state?" He had already advocated a state collection and scientific academy in Texas, but in the absence of one he wanted his work to be preserved elsewhere, in order that it might contribute to the growth of scientific knowledge.[11] When Buckley wrote Durand of his old friend's doubts, a letter came in April,

1860, from Dr. Thomas Stewardson, corresponding secretary of the Academy, assuring Gideon that their members were not "Black Republicans" and that they held him in high esteem despite his pro-slavery stand.[12] Satisfied but also somewhat embarrassed that his private fears about colleagues in the North had been made public, Gideon wrote Durand (January 13, 1861):

> You must not think amiss of me for the course I have taken . . . , for I am persuaded that, if you knew the infernal work, that has been perpetrated, and is still going on here, by those ignorant priest-driven fanatics of the Northern section of the Confederacy [i.e., Union], you would not blame me at all for being a little reluctant in doing favors [for] such inveterate enemies. With liberal minded gentlemen anything I can do in my small way that will add a single line to the record of the natural—the only true sciences,—shall be done as free as the water flows.

A letter Lincecum wrote Durand on March 12, 1861, comments significantly on his allegiance to Darwin and also shows that graphology was one of the pseudosciences he practiced:

> I have quite a number of never-seen correspondents, some of them natural and others again somewhat swelled. Some of them have the desire, but are too indolent to climb the steep hill of fame. Others again are all energy, and by the aid of pretty fair talent, and a good many favorable circumstance[s], have been elevated to a giddy height. Such is the condition of Mr. Charles Darwin, of Down Bromley Kent. Eng. Charles Darwin is a good man and an earnest seeker for natural, and scientific truth; has been eminently successful; and in his recent work on the subject of *natural selection* has shocked old science and theology exceedingly. Success to him, and I shall aid him, by proping him with all the *little sticks* I may possess that will answer the purpose. I am, to a considerable extent, his disciple; was a disciple of his grandfather Erasmus Darwin, during his lifetime. Yet, from the indications displayed in his autograph, I see, and he may not be cognizant of it, that Charles Darwin occupies a plane above his nature. I could name others who similarly placed, by fortuitous circumstances, are addicted to airs and over pretentions. But you, my esteemed friend, are not, in my catalogue, placed among any of the above orders, or species of scientific investigators.

Lincecum's last letter to Durand before the Civil War interrupted their correspondence was dated April 11, 1861, and it included a number of observations on ants:

I am daily employed in the investigation of Texas ants. I have already preserved in alcohol 34 distinct species, all found within 15 miles of my residence. They are such an extremely interesting genus that it will take a long time to satisfy myself in regard to their manners and customs. We have one species, who are agriculturalists; another species, who are horticulturalists, have slaves, and will excavate tunnels an inch in diameter to the distance of three or four hundred yards, sometimes passing it under smart streams of water. And these immense public works are accomplished by the labour of their slaves. They dwell in large and very ancient cities, on elevated sites, always in sandy lands, dig wells, sometimes very deep, 60 or 70 feet; and as they cannot stand the sun in hot weather, to obviate that inconvenience, they plant shade trees; or they could do no work in the summer daytime. All the species possess some traits of character peculiar to themselves, very interesting. Their wars are frightfully disastrous, particularly when it takes place between two kingdoms of the same type. I have seen, not less than a gallon—40,000—left on a single battle field. Few of them were dead; but the most of them had their legs all trimmed off, and they lay on the ground, amongst the scattered fragments of their dissevered limbs, doubling and gnashing their legless bodies in an agony of sullen, mad, hopeless dispair. But I must not write about ants. I set out to talk with you about other matters— that may suit you better, and take less time. And as I could not, in a month, tell you what I have seen the ants do, I'll say no more about them. . . . This *letter is not intended as a contribution*!!! Just for yourself.

By the time their correspondence picked up again, more than four years later, Lincecum had spent almost sixteen years studying Texas ants. In what appears to be the letter which renewed contact with Durand (dated December 24, 1865), Lincecum begins thus:

Your very kind, fraternal communication of 10th Novr 1861 reached me on the 11th. mar. 1862. It came open and represented, by the several endorsements on its back, that it had made its way through the lines of the contending parties, with many approvals. It had been respectfully handled and made its passage very safe and clean. You express much regret at the breaking up of our agreable, as you are pleased to call it, correspondence. No one could regret the abrupt manner in which our, as I think, quite profitable correspondence was cut short than myself. I tried, as you requested me to do, to get a letter through to you several times, but as I heard no more of you, I suppose they did not reach you, and now, I am not sure that you will ever see this. If however, you do get it, I claim an immediate answer.

I shall not be able to interest you any at this writing. Every thing is ruined here, so badly that it will require three or four years to repare and fix up our homes so we can spare time to investigate any of the sciences, except perhaps, those branches which are concerned in procuring the actual material for sustaining existance. What I could do was of but small importance in a scientific point of view, and that little has been ruthlessly suspended by the gross animal action of a half civilized race of speculator. I hope your time and spirit of investigation has not been so badly interrupted. I have clearly lost four full years of my already greatly extended period of existence.

I forget now how many species of Texas ants I had collected when I last wrote to you. I have 42 well preserved species now and it takes several quires of paper to tell what I have seen them do, in their civil, religious and political action, with their destructive wars, and their slavery system, which obtains almost universally. . . .

In 1865, Gideon sent his "histories" of these forty-two species to the Academy of Natural Sciences in Philadelphia. The accounts are quite interesting, often revealing Gideon's belief in the reasoning powers of these tiny creatures. In his account of "No. 8—Microscopic Red ant,"[13] he wrote:

The inginuity of this diminutive species of ant is truly wonderful. Small as he is, he can invade the habitation of man with almost resistless force, an[d] in some instances perpetuate extensive mischief. Dr. Carter, of Chappel Hill, had an infant, who was as the doctor supposed, afflicted with sore eyes, and notwithstanding the doctors constant visits, the child continued to grow worse every day. At length a neighbor lady being present, and who, having suspicions that all was not right in regard to the child's eyes, made a close examination when she discovered that the edges of the childs eyelids were thickly set with this species of ant, who were eating away the flesh, and that nothing else ailed the child. The lady cleared them all away, but they soon returned, and were working away at the childs eyes again. The little crib in which the child lay was now scalded out and the feet of it placed in basons of water. It was not many hours until they were gnawing away at the child's eyes as before. On examination it was ascertained that the persevering little pests had crossed, and were still crossing, the water in the basons at the foot of the crib, by a kind of pontoon bridge, made however, of ants. The manner was very singular and ingenious. They collected in considerable numbers at the waters edge, and after seeming to make observation, one of them would swim in, another, holding by the hind legs, would soon follow, and in the same manner, a third, and a fourth, and so on, till the for[e]most one of the series would reach the foot of the crib, and immediately begin climbing up, then another and another, and all the time great numbers would run over on the backs of those who were swimming over in the chain series.

When he responded to Durand's request for more details on ant battles with a vivid and colorful account of a conflict between two kingdoms of his "large black tree ant," Durand shared the letter with his friend and fellow member of the Philadelphia Academy, Joseph Leidy, who read it before the Academy.[14] Most of those present considered it fanciful rather than factual, but Leidy and Durand defended Lincecum. When told of this contretemps, Lincecum replied in a letter to Durand (dated June 16, 1866):

> Please present me respectfully to Prof. J. Leidy and tender to him my sincere thanks for defending the little history of the ant battle which I had attempted. . . . Please say to him that we must not entertain any feeling on account of the unbelief of these men, as they perhaps, have seen only the dry specimens of the ants in question, and therefore could not know anything in relation to their living character and belligerent habits. These cabinet investigators have seldom felt the tick bites and stings of the warrior ants in the forests and plains traversed by the industrious and earnest examiner of nature. Let them try the experiment of going out and looking for themselves, and if they seek earnestly, they will find many things true which their present mental condition would stile poetical and imaginative. They would find that ants do have wars, and that twenty or forty thousand of some species of them, engaged on a single battlefield, is nothing for the initiated to wonder at. . . .

Despite the skepticism of those who heard it read, a portion of Lincecum's account of the ant battle was published in the Academy's *Proceedings* for January, 1866:[15]

> Dr. Leidy read several extracts from a letter of Dr. Gideon Lincecum, addressed to Mr. Durand, dated Long Point, Texas, Dec. 24, 1865. One of the extracts related an interesting account of an ant battle, witnessed by Dr. Lincecum, as follows: "The large, black tree ants have exceedingly destructive wars sometimes with their own species. Like the honey bee, they maintain separate and distinct governments, or hives, and between these, as far as my observation goes, there is no commerce or intercourse of any description. But they have territorial claims and quarrels; and these quarrels are occasionally decided on

the battle field. As they are equal in physical strength and the science of war, the amount of life that is destroyed in one of their national conflicts is sometimes very great. I have seen left on one of their battle fields at least a gallon of the slain. They were not dead but they were in a far more lamentable condition. Their legs having been all trimmed off, they lay on the ground amongst the scattered fragments of their dissevered limbs, wallowing and writhing their legless bodies, in an agony of sullen, mad, hopeless despair.

This disastrous engagement took place in the little front yard of my office, on the evening of the 10th of July, 1855. There were considerable numbers engaged in battle when I first observed them. They were madly fighting in a hand to hand conflict, and reinforcements were momentarily arriving to both armies. The battle had now become general, and was raging over an area of 15 to 20 feet in diameter. It was 4 P.M., and placing a chair in a convenient situation for observation, I seated myself, for the purpose, if possible, of ascertaining the cause of the difficulty, and to note their mode of warfare. I was not present at the commencing of the battle, and now, while it was wildly raging, could not find out the cause of it. It was not long, however, until I disovered that the belligerent parties were the subjects of two neighboring kingdoms, or hives, each of which, as I could distinguish, by the arrival of their reinforcements, were coming from two different post-oak trees, which were standing about fifty yards apart, and the office-yard being very nearly the half-way ground, afforded me good opportunity to determine that the contending parties belonged to distinct communities, and not to the same hive.

The battle continued unabated, until the darkness of the night prevented further observation. I left them to their fate, with my feelings so highly excited that I did not rest well that night. Before sunrise the next morning I visited the battle field and found it thickly strewed with the legless, hapless warriors, as described above. There could not have been less than 40,000 left on the ground who were utterly incapacitated to help themselves. A few of them had a single leg left. With this they made shift to pull themselves incessantly around in a very limited circle. The larger proportion of them lay prostrate, writhing and doubling, and vainly straining their agonized, limbless bodies in a state of mental abandonment and furious desperation. Few were dead. All the dead ones that I saw, did not exceed perhaps a hundred; and these were found universally in pairs, mutually grappling each other by the throat. With a few of these pairs of unyielding warriors, life was not entirely extinct. My sympathies being painfully excited, I made an effort, where there were signs of vitality, to separate them. In this I did not succeed. On closer scrutiny I found that they had fixed their

caliper-like mandibles in each others throat, and were gripped together with such inveterate malignity, that they could not be sepa-
rated without tearing off their heads.

I had swept them up in a heap, and as the most humane method
of curtailing the wretched condition of the poor, ruined victims of the
bloody strife I could think of, was making a hole in the ground, with
the intention of entombing the whole of them, Whig and Tory together, and by filling the grave with water, drown them. But before I
had completed my arrangements, there came a heavy shower of rain,
which soon overwhelmed them with mud and water, thereby reliev-
ing me from the painful task.

It is perhaps nothing amiss to state here, that among the slain—
the vanquished—I saw no type of the species, except the neutrals, or
working type. As on the ensanguined fields of the arrogant genus
homo, the conjuring priests and better bloods of the self-created no-
bility, after raising the *fuss* had found it convenient to have business in
some safer quarter. . . .

Despite their colleagues' skepticism, Durand and Leidy encouraged
Lincecum to send the Academy more reports on his observations of
ants. On February 22, 1866, Lincecum sent Durand a letter which included "Sketches of his [the small, black, erratic ant] history, a portion
of which was published in the *Houston Telegraph*. . . . I have clipped from
the *Telegraph*, the portion that was printed and shall enclose it, and the
balance of the manuscript herewith. It will, if nothing more, . . . amuse
you a little." Durand obviously considered Lincecum's "sketches"
significant. The March/April, 1866, *Proceedings* include the following entry dated April 10th:[16] "Mr. Vaux, Vice President, in the Chair. Twenty-
nine members present. A letter was read from Dr. G. Lincecum of Texas,
containing a history of the 'small black erratic ant.'"

This report was quite lengthy, amounting to six pages of small
type, single-spaced. The opening portion and another battle description are worth quoting:

The small, black, crooked running ant, so common in everybody's
yard, and on almost every growing twig in spring time and summer, is
called, in my catalogue of ant species, the erratic, or crazy ant. He is
No. 5 in my notes on the various types of ants.[17] In this species the
formic acid odor is very strong when the ant is crushed. He is quick in
his movements, does not make paths, but travels in scattered files, in
the same direction, sometimes several hundred yards; moves quickly
on a general course, running very crooked the whole route, giving his
path a broad range, travelling two or three times the distance to his
place of destination. All along the range of their path, at unequal dis-

tances, are depots or station-houses, at which they often call as they pass along, giving the whole affair quite a business aspect. . . . Cripple one of them on the route of his travel, and you produce the wildest excitement, and the invalid will be visited and examined by perhaps 500 of the traveling throng in the course of two or three minutes. If the case is a curable one, they work with him until he is on foot again, when he moves onward with the crowd as before. If he dies, they remove him from the range of the great thoroughfare, and business rolls on again.

They sometimes wage war with the red-headed tree ant, . . . and the conflict is generally quite disastrous. Notwithstanding the fact that they are always able to bring to the field more than ten times the number of their red-headed foe, they often meet with defeat. I was spectator to a battle, or rather a field fight, between these two species of ant, that continued four or five hours. Small parties were engaged in the deathly conflict at sunrise, when I first observed them. They were fighting in the wagon road, and their numbers were rapidly increasing. At the time I was called to breakfast, they were in considerable force on both sides, and when I returned I found both armies greatly augmented. Reinforcements were constantly arriving, and the battle was raging over an area of eight to ten feet in diameter. The discipline and modes of battle of the two species are entirely different. The method of attack, by the little black ant, is aimed altogether at the feet and legs of the foe; and as they greatly outnumber the red heads, by engaging them two or three to one, they succeed in maiming and rendering large numbers of them unfit for service. The red heads seem to aim only at decapitation, and this they accomplish with dexterity and surprising facility. Reinforcements were momentarily arriving to both armies. Thousands were already engaged, and the bloody strife was raging over the entire area of the battle-field.

Being controlled only by two forces,—desperation and death— the scene was terrific beyond my powers of description. In all directions, everywhere, was seen the dire effects of relentless war. The battle-field was already thickly strewn with the dead and dying, over whom, in regardless tramp, swept the furious antagonism. Here indeed was, for once, at least, full manifestations of the unmistakable, genuine "tug of war." Violently struggling and gnashing their jaws, clinging together and wallowing on the ground, in companies, in squads and single combat, the direful contest fiercely raged. Dispatches had been sent off by the black ants for their entire reserve to be forwarded immediately, and they were pouring out by the million from the gates of their great city,—distant about 60 feet,—and hurrying toward the battle-field. They were evidently making a forced march, and their numbers were so great that by the time they had progressed 20 to 30 feet, their line of march suggested the idea of a broad black ribband trailing on the ground, and there seemed to be no end to them, for they were still flowing out from the city in countless thousands.

At this crisis their army on the battle-field gave way and was routed, and in a general panic commenced a retreat. Soon, in their disorderly flight, they met their reinforcements and communicating to the front ranks their total and disastrous discomfiture, the panic became universal, and reinforcements and all fled precipitately into the city. In five minutes there were no black ants to be seen above ground.

In response to the request for a detailed report on his agricultural ant, Lincecum wrote nineteen pages, condensing the history as much as possible, but protesting that it would require a volume to do them justice. After he and Leidy read it together, Durand reported that Leidy would present it, but he issued a discreet warning (which Lincecum later quoted in a letter to H. C. Wood, Jr., dated December 6, 1866):

> . . . my dear Lincecum, I am really afraid that some will prove unbe-lievers in such wisdom as you attribute to this little insect. People are so accustomed to see *but one* being in creation endowed with reason that *they* will look upon you as *one possessing great imaginative powers* and that *you make too much* of what is called animal instinct. *They* will not exactly doubt what you relate; but they will be persuaded that you lend to your narative *your own reason*, your own perspicacity and that all you imagine seeing is nothing but illusion. . . . All that you say of the agricultural ant is certainly very wonderful, and it must be in *your mind* exactly what you say.

Just as Durand had predicted, Gideon's account was once again viewed with skepticism by most of those who heard it read, but it was ac-cepted for publication in the *Proceedings*, under the title "On the Agricul-tural Ant of Texas."[18] We have chosen to reprint the bulk of it here because it extends Gideon's observations beyond what he reported in the letters to Charles Darwin and, more importantly, gives the earliest account of a founding of a new colony by a single queen.

> This is No. 2 of my catalogue—is inodorous, having no smell of formic acid. It is a large reddish brown ant, dwells in the ground, is a farmer, lives in communities which are often very populous and controlled by a perfect government; there are no idlers amongst them. They build paved cities, construct roads, and sustain a large military force.
>
> When one of the young queens, or mother ants, comes to matu-rity and has received the embraces of the male ant, who immediately

dies, she goes out alone, selects a location and goes rapidly to work excavating a hole in the ground, digging and carrying out the dirt with her mouth. As soon as she has progressed far enough for her wings to strike against the sides of the hole, she deliberately cuts them off. She now, without further obstruction, continues to deepen the hole to the depth of 6 or 7 inches, when she widens the bottom of it into a suitable cell for depositing her eggs and nurturing the young. She continues to labor out-doors and in until she has raised to maturity 20 to 30 workers, when her labor ceases, and she remains in the cells supplying the eggs for coming millions, and her kingdom has commenced. But very few of the thousands of mother ants that swarm out from the different kingdoms two or three times a year succeed in establishing a city. However, when one does succeed in rearing a sufficient number of workers to carry on the business, she entrusts the management of the national works to them and is seen no more outside.

The workers all seem to understand the duties assigned to them, and will perform them or die in the effort. The workers increase the concealment, which had been kept up by the mother ant during the period of her personal labors, of the passage or gateway to their city, by dragging up and covering it with bits of stick, straw and the hard black pellets of earth, which are thrown up by the earth worms, until there is no way visible for them to enter; and the little litter is so ingeniously placed, that it has more the appearance of having been drifted together by the wind than to have been the work of design. In about a year and a half, when the numbers of the community have greatly increased, and they feel able to sustain themselves among the surrounding nations, they throw off their concealment, clear away the grass, herbage and other litter to the distance of 3 or 4 feet around the entrance to their city, construct a pavement, organize an efficient police, and, thus established, proclaim themselves an independent city. The pavement, which is alway kept very clean, consists of a pretty hard crust about half an inch thick and is formed by selecting and laying such grits and particles of sand as will fit closely over the entire surface. This is the case in sandy soil, where they can procure coarse sand and grit for the purpose, but in the black prairie soil, where there is no sand, they construct the pavement by levelling and smoothing the surface and suffering it to bake in the sunshine, when it becomes very hard and firm. That both forms of these pavements are the work of a well planned design, there can be no doubt with the careful investigator. All the communities of this species select their homes in the open sunshine, and construct pavements. Their pavements are always circular and constructed pretty much on the same plan. . . .

The mound itself, and the surface of the ground around it, to the distance of four or five feet, sometimes more, from the center, is kept very clean, *like a pavement*. Everything that happens to be dropped upon the pavement is cut to pieces and carried away. The largest dropping from the cows will in a short time be removed. I have placed a large corn-stalk on the pavement, and in the course of two or three days found it hollowed out to a mere shell; that too, in a short time, would be cut to pieces and carried off. Not a green thing is suffered to grow on the pavement, with the exception of a single species of grain-bearing grass, *(Aristida stricta)*. This the ant nurses and cultivates with great care, having it in a circle around and two or three feet from the center of the mound. It also clears away the weeds and other grasses all around outside of the circular roll of *Aristida*, to the distance of one or two feet. The cultivated grass flourishes luxuriantly, producing a heavy crop of small white, flinty grains, which, under the microscope, have the appearance of the rice of commerce. When it is ripe it is harvested by the workers and carried, chaff and all, into the granary cells, where it is divested of the chaff, which is immediately taken out and thrown beyond the limits of the pavement always to the lee side. The clean grain is carefully stored away in dry cells. These cells are so constructed that water cannot reach them, except in long wet spells, when the earth becomes thoroughly saturated and dissolves the cement with which the granary cells are made tight. This is a great calamity, and if rain continues a few days it will drown out the entire community. In cases, however, where it has continued long enough only to wet and swell their grain, as soon as a sunny day occurs they take it all out, and spreading it in a clean place, after it has sunned a day or two, or is fully dry, they take it in again, except the grains that are sprouted; these they invariably leave out. I have seen at least a quart of sprouted seeds left out at one place.

They also collect the grain from several other species of grass, as well as seed from many kinds of herbaceous plants. They like almost any kind of seeds—red pepper seeds seem to be a favorite with them. In a barren rocky place in a wheat field, a few days after harvest, I saw quite a number of wheat grains scattered over the pavement of an ant city, and the laborers were still bringing it out. I found the wheat quite sound, but a little swelled. In the evening of the same day I passed there again; the wheat had dried, and they were busily engaged carrying it in again.

The species of grass they so carefully cultivate is a biennial. They sow it in time for the autumnal rains to bring it up. Accordingly, about the first of November, if the fall has been seasonable, a beautiful green row of the *ant rice*, about 4 inches wide, is seen springing up on the pavement, in a circle of 14 to 15 feet in circumference. In the vicinity of this circular row of grass they do not permit a single spire of any other grass or weed to remain a day, leaving the *Aristida* untouched until it is ripe, which occurs in June of the next year, they gather the seeds and carry them into the granaries as before stated. There can be no doubt of the fact that this peculiar species of grass is intentionally planted and, in farmer-like manner, carefully divested of all other grasses and weeds during the time of its growth. And that after it has matured and the grain stored away, they cut away the dry stubble and remove it from the pavement, leaving it unencumbered until the ensuing autumn, when the same species of grass, and in the same circle, appears again, receiving the same agricultural care as did the previous crop; and so on, year after year. . . . This species of ant subsists entirely on vegetable seeds. I have sometimes seen them drag a catterpillar or a crippled grasshopper into their hole, that had been thrown upon the pavement, but I have never observed them carrying any such things home that they had captured themselves. I do not think they eat much animal food.

I have often seen them have prisoners, always of their own species. I could not discover the nature of the offence that led to the arrestment; still I have no doubt as to the fact of its being so and that the prisoner is very roughly forced along contrary to its inclination. There is never more than a single guard having charge of a prisoner, who by some means having obtained the advantage, and attacking from behind, had succeeded in seizing it with the mandibles over the smallest part of its back, and so long as it maintains this grip it is out of the reach of harm from the prisoner.

In some cases the prisoner quietly submits, and folding up its legs, forces the captor to carry it along like a dead ant, as I thought it really was, until I caused its captor to drop it; when, to my surprise, it immediately sprang to its feet, and, running wildly, succeeded in making its escape. It occurs more frequently, however, that the prisoner does not give up so tamely, but continues to make every effort to rid itself of its detainer. I have many times observed the prisoner manifesting all the indications of terror and great reluctance at being so unceremoniously dragged along. It will lay hold of and cling to everything that comes in reach, and by this means greatly retard the progress of its captor. When at last they arrive on the city pavement, half a dozen or more of the national guard, who are always on duty, rush upon the prisoner, aiding the seemingly fatigued captor, who still maintains its potent grip upon the now almost helpless prisoner, seize it by the arms, legs, everywhere, and in a very rough manner hurry it down into the entrance to the city, and out of the reach of further observation.

The agricultural ant is very tenacious of life. I dissevered the head of one at 4 P.M. on Sunday, and the head remained alive, retaining sufficient strength by pressing with its antennae against the slip of glass upon which it lay to move itself and change its position, until 10 A.M. the next day.

It seems to be an established law amongst all species of ants, and particularly with the species in question, that when any disaster occurs to their city, the first thing to be done is to take care of the young, and, if possible, secure their safety; and so, when by any accident one of their cities gets torn up, it will be seen that they universally rush to the nursery apartment; and every one that can, takes up an egg, the pupae, the young in any stage of advancement, and will save its life or lose its own. As far as I can understand and read their actions, every one understands its duty, and will do it or lose its life. I have observed the guards, when a sudden shower of rain would come up, run to the entrance of the city, and there meeting with another party coming up from below, would crowd themselves together in the hole in such manner as to form a complete obstruction to the ingress of the water, and there remain overwhelmed with the accumulating rain until it ceased. If the shower continues over fifteen minutes, they are found to be still closely wedged in the aperture and all dead; and there they remain until the balance of the pavement guards, who during the shower had climbed some weed or blade of grass that grew near the border of the pavement, come down, and with some difficulty succeed in taking them out. They are immediately taken to some dry place on the pavement and exposed to the open air half an hour at least; after which, if they do not revive, they are taken off from the pavement, sometimes to the distance of sixty yards, and left on the ground without further care.

Long-continued rainy seasons, by deeply saturating the earth, will dissolve the cement of their cells, flood them, and drown the ants out entirely. I have allusion now only to the agricultural species of the genus. The first year after my arrival in Texas, I noticed that there were a great many uninhabited ant hills, with pavements still smooth and nude of grass or weeds, indicating that they had been very recently occupied. The missing communities were all dead—extinct—had been destroyed by a series of rainy seasons. Then, there were but few of these ant cities to be found that were occupied. But when the drouth set in, the earth being no longer filled with water, they began to mul-

tiply very rapidly. City after city appeared as the dry weather continued, and now, 1863, at the close of a ten years' drouth, they have spread
so extensively that their clean little paved cities are to be seen every
fifty or sixty yards, especially along the roadsides, in the prairies, walks
in yards and fields, barren rocky places, &c. In beds of heavy grass, or
weeds, or in deep shady woodlands, they very seldom locate a city.
They prefer sunshine and a clear sky. This ant does not work in the
heat of the day during hot weather but makes up the lost time during
the night. I have often found them busily engaged at two and even
three o'clock, A.M. Before day, however, they call off the workers, and
rest till about sunrise. In more favorable weather, when they can op-
erate all day, they do not work late at night.

In regard to courage, there can be no mistake in stating that when
the interests of the nation are involved, this ant exhibits no signs of
fear or dread of any consequences that may result to self while engaged in the discharge of its duties. The police or national guards of a
community which has been established three or four years number,
in the aggregate of the parties on duty, from one to two hundred.
These are seen all the time, in suitable weather, unceasingly prom-
enading the environs of the city. If an observer takes his stand near
the edge of the pavement, he will discover an instantaneous move-
ment in the entire police corps, coming wave-like towards him. If the
observer imprudently keeps his position, he will soon see numbers
of them at his feet, and without the slightest degree of precaution, or
the least hesitation, they climb up his boots, on his clothes, and as
soon as they come to anything that they can bite or sting, whether it
be boot, or cloth, or skin, they go right to work biting and stinging;
and very often, if they get good hold on any soft texture, they will
suffer themselves to be torn to pieces before they will relinquish it. If
they succeed in getting to the bare skin, they inflict a painful wound,
the irritation, swelling and soreness of which will not subside in
twenty-four hours.

If any worm or small bug shall attempt to travel across their
pavement, it is immediately arrested and soon covered with the fear-
less warriors, who in a short time deprive it of life. Woe unto any
luckless wight of a tumble-bug who may attempt to roll his spherical
treasure upon that sacred and forbidden pavement. As soon as the
dark, execrable globe of unholy material is discovered by the police
to be rolling on and contaminating the interdicted grounds, they rush
with one accord upon the vile intruder and, instantly seizing him by
every leg and foot, dispatch him in a short time. Sometimes the tumble-
bug takes the alarm at the start, while only two or three of the ants
have hold on it, expands its wings and flies off with them hanging to

its legs. If it fails to make this early effort, it very soon falls a victim to the exasperated soldiery. The ball of filth is left on the pavement, sometimes in the very entrance to the city. In due time the workers take possession of it, cut it into fragments, and pack it off beyond the limits of the incorporated grounds.

I have not observed that anything preys to any considerable extent upon this species of ant. Chickens and mocking birds will sometimes pick up a few of them, but not often. If anything else in Texas eats them, I have not noticed it. Neither have I observed their nests bored into or dug up in middle Texas. The agricultural ant is of but little disadvantage to the farmer, however numerous, as it is never seen six inches from the ground, nor does it cut or trouble any growing vegetable outside of its pavement, except the seeds of the noxious weeds and grasses. Sometimes it is found stealing corn meal, broomcorn seeds, &c.; but it is only when it finds them on the ground that it steals even these.

Children occasionally get on their pavement, and are badly stung. A few of these pavement lessons, however, generally obviate that inconvenience. The pain of their poison is more lasting, will swell and feel harder, than that of the honey bee. If they insert their stings on the feet or ankles of the child, the irritation will ascend to the glands of the inguinal region, producing tumours of a character quite painful, often exciting considerable fever in the general system; the irritation will last a day or two, but I have seen no permanent injury arising from it.

During protracted spells of dry weather, they are frequently found in great numbers in our wells. They seem to have gone there in pursuit of water, and not being able to get back, to make the best of a bad condition—in this unforeseen dilemma—they will collect and cling together in masses as large as an ordinary teacup, in which condition they are frequently caught and drawn up in the bucket. When they are thus brought up, though they may have been in the water a day or more, they are all living, though half drowned and barely able to move. While in the well they are all afloat, and at least one-half the mass submerged. As it is known that this species of ant cannot survive 15 minutes under water, how they manage when in a large half-sunken mass to survive a day, or even longer, is a question to which I may fail to give a satisfactory solution. I may, however, from experiments I have made with single individuals in water, venture the assertion that there is no possible chance for the submerged portion of the globular mass, if it remain in the same condition in relation to the water, to survive even half an hour. Then we are forced to the supposition that by some means or other the ball must be caused to revolve as it floats. The globular mass must be kept rolling, and make a revolution every four minutes, or the submerged portion must die. To accomplish this somewhat astonishing life-preserving process, there is but one possible alternative. It can be effected only by a united and properly directed systematic motion of the disengaged limbs of the outer tier of ants, occupying the submerged half of the globular mass.

I saw to-day (June 15), in a clean-trodden path near my dwelling, quite a number of this species of ant engaged in deadly conflict. They were strewed along the path to the distance of 10 or 12 feet, fighting, most of them, in single combat. In some few cases, I noticed there would be two to one engaged, in all of which cases the struggle was soon ended. Their mode of warfare is decapitation, and in all cases where there were two to one engaged the work of cutting off the head was soon accomplished. There were already a number of heads and headless ants laying around, and there was a greater number of single pairs of the insatiate warriors grappling each other by the throat on the battle-field, some of whom seemed to be already dead, still clinging together by their throats. Among the single pairs in the deadly strife there were no cases of decapitation. They mutually grapple each other by the throat, and there cling until death ends the conflict but does not separate them. I do not think that in single combat they possess the power to dissever the head; but they can grip the neck so firmly as to stop circulation, and hold on until death ensues without their unlocking the jaws even then.

The cause of this war was attributable to the settlement of a young queen in close proximity (not more than 20 feet) of a very populous community that had occupied that scope of territory for ten or twelve years. At first, and so long as they operated under concealment, the old community did not molest them; but when they threw off their mask and commenced paving their city, the older occupants of that district or territory declared war against them and waged it to extermination. The war was declared by the old settlers, and the object was to drive out the new ones or exterminate them. But the warriors of this species of ant are not to be driven. Where they select a location for a home, nothing but annihilation can get them away. So, in the present case, the war continued two days and nights, and resulted in the total extermination of the intruding colony. From the vastly superior numbers of the older settlers, though many of them were slain during the war, they nevertheless succeeded in destroying the entire colony, without any apparent disturbance or unusual excitement about the great city. Their national works and governmental affairs went on in their ordinary course, while the work of death was being accomplished by their resolute bands of triumphant warriors. . . .

The extensive, clean, smooth roads that are constructed by the agricultural ants are worthy of being noticed. At this season of the year their roads are plainest and in the best order, because it is harvest time, and their whole force is out collecting grain for winter supplies.

I am just this moment in from a survey of one of these roads, that I might be able to make an exact and correct statement of it. It is over a hundred yards in length, goes through twenty yards of thick weeds, underruns heavy beds of crop grass 60 yards, and then through the weeds growing in the locks of a heavy rail fence 20 yards more; and throughout the whole extent it is very smooth and even, varying from a straight line enough, perhaps, to lose 10 or 12 yards of the distance in travelling to the outer terminus. It is from 2 to 2 1/2 inches wide; in some places, on account of insurmountable obstructions, it separates into two or three trails of an inch in width, coming together again after passing the obstruction. This is the main trunk, and it does not branch until it crosses the before-named fence, beyond which is a heavy bed of grain bearing weeds and grass. Their prospecting corps travel far out, and when they discover rich districts of their proper food they report it, and a corps of foragers are immediately dispatched to collect and bring it in.

27th June, 1863.—My son, Dr. Leonidas, called my attention to an assemblage of the males and females of the agricultural ants (*Myrmica molefaciens*) which took place about 2 P.M., and continued in session until 4 P.M. They were all winged ants, and there were many thousands, perhaps millions, of them, thickly covering the ground over an area of 107 yards in length and 10 wide. They came from all directions and were evidently the production of many kingdoms of this wonderful species of ant. There must have been at least five males to one female, and all parties were rushing hither and thither over the entire area, described above, in a frantic, amative furor. Each female would be found covered and wallowing on the ground with clusters of from four or five to twenty males, and there were hundreds thickly rushing over the ground in search of females that were not to be found. The air was full of them flying around, going off and returning; some of them, perhaps, just arriving.

When a female became satisfied with her numerous lovers, by a great and violent effort she made shift to extricate herself from their rude embrace and immediately fly away. After 4 P.M. they began rapidly to fly away, and in the course of an hour they were all gone, leaving their disconsolate, exhausted lovers, who made no effort to follow. Many of the males were already dead, and a still greater number lay helpless on the ground; but there were hundreds of thousands who were still active, and they collected together in the horse tracks, cracks in the ground, and other places sheltered from the south wind, which prevails at that season of the year, and becoming perfectly quiet, were, at 6 P.M., lying still in heaps of from half a pint to a quart, sometimes more. At this hour I examined the entire field, and there must have been very near, if not quite, a bushel of the exhausted and dying male ants.

A strong south wind was blowing during the time the females were flying off, and the larger portion of them were drifted by the wind into the timbered lands to the north; many of them, however, succeeded in forcing their way a few hundred yards against the wind, and alighting, which seemed to be the effect of fatigue more than desire, they immediately, by writhing and doubling themselves in various ways, cast off their wings, which were no longer necessary, and running rapidly till they found a little clean spot of earth, went hurriedly to work digging holes in the ground, which they accomplished with apparent ease and considerable facility. They dig and bring out the dirt in considerable pellets with their large caliper-like mandibles, carrying it not exceeding two inches and dropping it in a circle around the hole they are making; very soon they had buried themselves out of sight. Two hours after they had commenced flying away from their lovers, hundreds of holes, with a little circle of black dirt around them, might be seen. In every clean-trodden piece of ground, and in the roads and paths, these new tenements were thickly set long before sun down.

Only one of these mother ants is necessary to start a kingdom. I saw no instance where two of them were at work at the same hole. In some favorable spot of ground there would be found a great many of them at work excavating their holes, sometimes within a foot of each other. None seemed to know that any other ant was near. While one was out with a load of dirt, I placed a stick in her hole; returning, she did not know the place, and in searching around soon found another one's hole, into which she immediately plunged. Very soon the owner of the establishment pushed the intruder out, who made battle as soon as they were fairly out on level ground. The conflict soon became desperate, and after they had fought for the space of a minute or two the intruder seemed to give way, and, extricating herself from her highly incensed antagonist, plunged into the hole again; the owner followed, and after some time succeeded in dragging the invader out once more, and also, after a dire conflict, in putting her to flight. The victor went to work again, but in the fight she had been injured, as I noticed every time she came out with a load of dirt she would stop awhile, and with one of her feet rub and fix something about her mouth. She seemed to be in pain, and did not work so vigorously as before the fight.

It would not do for many of these new queens to prove successful in building up kingdoms. There is some antagonistic action to prevent it. The male and female congress, I have attempted to describe above, happens two or three times every year, and should all the queens succeed in establishing colonies, they would in a very few years occupy the entire surface of the earth. This species of ant . . . do not go off from the old hive in swarms like the bee, but a single mother ant, after congress with the males, goes off alone and sets up for herself. She works very busily until she has raised 20 or 30 neuters to work for her, when she ceases to labor, and, remaining in-doors, lays all the eggs that produce the coming millions. The laborers are long-lived, so are the queens.

28th. — I extract from my journal: This morning I found the males where I left them last evening. The greater portion of them were still active, and seemed to be quite careless as to their fate. Hundreds were dead or dying. Great numbers had climbed up the little weeds, many of whom were dead, but still clinging by their jaws, which were fast gripped to some little leaf or twig. The females had buried themselves by the time it was dark last night, and, closing up their holes, remained shut in all night. But few of them had opened their doors and gone to work at an hour by sun this morning. The number of their holes is truly wonderful. I saw many places where there were at least fifty of their holes to the square rod, and northwardly they extended for miles. When these mother ants succeed in boring their holes to the depth of six or seven inches they close them up and employ themselves widening the bottom of them a little, forming small cells for the purpose, as I suppose, of making room for the deposition of their eggs. They do not, as I can discover, need any food yet. At 5 P.M. of this day I visited the place again and found the male ants all dead. They were drifted into the gullies by the winds into heaps, and thousands of them besides lay scattered over the ground. Some of the females were still engaged deepening their holes, and their little piles of black dirt were to be seen everywhere.

29th. July. — A month has passed. I went round to-day and found that, in all those thousands of female ants, who made so brave a start excavating new homes, there was but one that was a success, and it was concealed with a little pile of trash. There may be more, but I did not find them, and the winds have swept away their little piles of dirt, so that there are no signs of them left. From some cause they are all gone. Eight or ten days after they had shut up their holes I dug up quite a number of them; found them looking well, but they had no eggs or anything else in the little cell. They seemed to be sleeping. I have never witnessed similar assemblages in any other species of ant, though I have seen it often take place with the agricultural species. *Long Point, Texas, Oct., 1866.*

It was this paper, more than any other, which prompted Henry Christopher McCook, a young naturalist and a Presbyterian minister in Philadelphia, to undertake a journey to Texas in 1877 that eventually led him to publish in 1879 a book entitled *The Natural History of the Agricultural Ant of Texas: A Monograph on the Habits, Architecture, and Structure of Pogonomyrmex Barbatus.* McCook spent three weeks observing agricultural ants near Austin, having chosen that site in part because it was within the range of Lincecum's observations. His first chapter states succinctly the history of the controversy which brought him to Texas:

> Mr. Moggridge,[19] in his admirable notes on harvesting ants, calls attention to "the principal questions which still await solution" in the life-history of those remarkable insects. . . . He places among the known harvesting ants the subject of this sketch, and cites correctly the authority by whom it became generally known to natural history, Dr. Gideon Lincecum, late of Long Point, Texas. This gentleman's observations were originally committed to Mr. Charles Darwin, by whom they were communicated to the Journal of the Linnaean Society of London, in a note read April, 1861. Five years afterwards, a paper was published in the *Proceedings of the Academy of Natural Sciences*, from the manuscript of Dr. Lincecum, which gave, with some detail, the result of observations upon this ant, extending throughout a long residence in Texas. Prior to either of the above dates, however, October 23, 1860, a paper was submitted to the Philadelphia Academy of Natural Sciences, entitled "The Stinging or Mound-making Ant, *Myrmica (Atta) molifaciens*, by S. B. Buckley." This paper was printed in the *Proceedings* (xii., 1860, p. 445), and is undoubtedly the first publication of the interesting habits of this emmet.
>
> Mr. Buckley's notes are incomplete, in some respects incorrect, but in the main are true to the facts as far as they go. Some of these facts are credited to Dr. Lincecum, and it is probable that to Mr. Buckley's interest in the matter we are indebted for that gentleman's paper. In view of the above facts, it would seem to be a strange oversight in one who had given so much attention to the habits of harvesting ants as Mr. Moggridge, to raise the question whether such creatures inhabit the Southern States of North America.
>
> Dr. Auguste Forel, in the bibliography accompanying his *Swiss Ants*,[20] gives expression to the same doubt, in connection with his references to Lincecum's notes. "These observations," he says, "inspire me

with little confidence." The doubt which is thus raised concerning Dr. Lincecum's observations is a fair index of the state of mind which I found to exist among the older members of the Philadelphia Academy, who had more or less knowledge of the author, and the origin of the paper above referred to. While it was believed that there was some basis of fact in the communications made, they were thought to contain much more that was fanciful, and indeed, a shadow of doubt rested upon the whole. In the course of my inquiries I found that the original manuscripts of this paper, as well as of one upon the Cutting Ant, the "erratic ant," and other Texan Formicariae, were in the hands of Mr. E. T. Cresson. They had been forwarded to him several years before, as an officer of the American Entomological Society. Mr. Cresson kindly placed these manuscripts in my hands. They were carefully read, and the reason for the suspicion with which they had been viewed was everywhere quite manifest. The venerable writer had many peculiar notions about society, religion, and the genus homo generally, which he could not refrain from thrusting—in the most untimely manner and objectionable words—into the midst of his notes. These idiosyncrasies, together with some peculiarities of spelling, grammar, and rhetoric more original than regular, had evidently raised in the minds of officers and members of the Academy a question, not as to the integrity of the author, but as to his accuracy as an observer. Several of his papers were, however (after proper corrections and condensations), published, among them the notes upon the agricultural ant. The unpublished papers in my hands have been freely used in the preparation of this work, and have contributed some valuable facts. In the summer of 1877 I was able to visit Texas for the purpose of settling, if possible, the questions which had been raised as to the accuracy of the reports of Buckley and Lincecum. As will be seen, the observations of Dr. Lincecum were, in many important points, confirmed during that visit, and thus a strong degree of authenticity given to other *facts* recorded by him that I was not so fortunate as to note.[21]

It is hardly surprising that Lincecum's work on the agricultural ant *(Pogonomyrmex molefaciens)* and other ants was suspect in many scientists' minds. Being mainly self-educated and a self-trained scientist, he had a tendency to personify his observations of these organisms and attribute "reason" to them.[22] For his unscientific approach, Lincecum received a good deal of scorn at the time of his publications as well as later. Consider two separate aspects of his work that received scrutiny: the supposed "planting" of the ant rice by the agricultural ants, which was eventually labeled "the Lincecum myth," and his detailed description of the establishment of a new colony by an ant queen. Both were discussed in detail in the classic text, William M. Wheeler's *Ants: Their Structure, Development, and Behavior* (1910),[23] and they deserve more consideration here.

The first of these refers to Lincecum's suggestion that the agricultural ant intentionally planted grain around the periphery of the mound. Wheeler first discusses how this idea took hold and then devotes two pages to refuting what he labels "the Lincecum myth": "This notion, which even the Texas schoolboy has come to regard as a joke, has been widely cited, largely because Darwin stood sponsor for its publication in the *Journal of the Linnean Society*. McCook, after spending a few weeks in Texas observing *P. molefaciens* and recording his observations in a book of 310 pages . . . failed to obtain any evidence either for or against the Lincecum myth. He merely succeeded in extending its vogue by admitting its plausibility." Wheeler then goes on to suggest an alternative explanation for the circle of grass around the ant mounds: that while the grains are deposited by the ants, the "sowing" is simply a discarding in an area away from the mound. In his words, it is "an unintentional and inconstant by-product of the activities of an ant colony."[24]

Now consider the second item, Lincecum's notable contribution in the form of an early account of the founding of a colony. In view of Wheeler's attack on "the Lincecum myth," it is easy to see why he stated that "the actual founding of a colony by a single queen was first witnessed by an American of somewhat doubtful reputation as a myrmecologist, Dr. Gideon Lincecum." He goes on, however, to note: "Essentially the same account is repeated in McCook's larger work on the Texan agricultural ant." Furthermore, when Wheeler gives his own account of "an unusually striking case of colony formation by queens of the California harvester," even he admits, "This recalls the above cited observation of Lincecum on the Texas harvester."[25] In the final analysis, Wheeler credits Lincecum with an early and accurate account of this important but little-known phenomenon.

In Bert Hölldobler and Edward O. Wilson's *The Ants* (1990), Lincecum's "famous misconception" is highlighted, along with Wheeler's retort. Hölldobler and Wilson report that after intense research on the harvester (agricultural) ants from 1860 to 1910, very little was done until the early 1970s. Recently, however, these ants have been the subject of both field and experimental studies on foraging and competition. For Lincecum the most important thing would be the fact that "his" ants are still being studied.

Chapter 2

Texas Botany: Gideon's Catalogue of Useful Plants

*What could afford the lover of nature so much pleasure, as to be able to
stroll out into the dark shady forests, at a leisure time, and converse
with every tree and shrub, twining vine, tender herb and blade of grass
he may chance to meet with?*

Gideon Lincecum

If Lincecum's ants were "sowers of grain," he was no mean sower himself. He had learned to farm from his pioneering father, but his interest
in nature extended well past merely raising a crop. It reached across
the whole realm of botany—and beyond. In a letter to Dr. George Engelmann of St. Louis, Missouri (November 25, 1867), Lincecum writes,
"My fondness for nature and natural things, originated, as I suppose,
from the fact, that I was raised, nearly to manhood, in the wild forests,
with the Muscogee Indians." His excursion with the *Alikchi chito* (Choctaw
doctor of great reputation) in Mississippi in 1832 gave him a practical
introduction to the usefulness of plants and their medicinal uses. This
six-week trip is detailed in Lincecum's autobiographical writings.[1] His
autobiography also indicates that he studied books on botany while
riding horseback to make calls on patients.[2]

Some thirty years later, in a letter dated September 13, 1865, to "Sioux"
Doran, Lincecum wrote:

> I have got in all the kinds of medicines that are to be found in this
> portion of Texas. The number is quite small of the kinds that are col-
> lected in autumn, and they are such as are applicable to winter com-
> plaints, Pleurisy, rheumatism, colds, &c. Nature never intended this

portion of Texas for a sickly country, there are so few medicines, that important cases are not indicated. In the spring, when everything is blooming—when singing birds, gaudy butterflies and painted flowers are filling the air and the landscapes with melody, gay sights and sweet odors—is the time to go forth into natures well furnished laboratory and help oneself to her greatest and best bounties for the sick man.

Also in a letter to R. B. Hannay, of Millican, Texas, January 24, 1864, Lincecum wrote:

Our prairies abound with precious medicines for the complaints of this climate. There is no good sense in going off to other climates and countries in search of medicine for this. All that is needed to counteract the different forms of disease belonging to our country can be found here. Nature never committed the foolish blunder of sowing the seed of the plants she intended as medicine to people of one country, in another. We recognize a general fitness of things everywhere else, and if we fail to find it where it is most looked for, and where it is most needed, it would be a manifestation of a very great and damaging defect in the plans of the world makers. There is no such deficiency, however—"the flora of every country is sufficient to meet the indications of the diseases thereof," said Dr. Barton,[3] and though I did not believe it, when he made the declaration 40 years ago, I have since had ample opportunity to demonstrate the truth of his declaration in many instances. I have moved southward, from latitude to latitude, and always found that the aboriginal inhabitant[s] could supply themselves from the flora of their own districts—though different from the plants in other latitudes—with all the medicines required for their complaints.

Beyond recognizing the medicinal plants, Lincecum knew the poisonous plants and included a discourse on these plants in the *Texas Almanac* of 1861.[4] The source of the *Almanac* article is a letter Lincecum wrote to the *Galveston Weekly News* (April 1, 1860) about the poisonous plants of Texas:

At present, I will speak of plants, and the indications by which almost any person may soon learn to avoid the plants which are poisonous.

Persons of all discriptions have frequent occasion to make some use of plants, and sometimes when they are not in a situation minutely to investigate their nature & qualities.

The following rules for extemperanous examinations will be found as a general rule, for avoiding poisons. These rules are the results of many investigating minds:

Plants with a glume calyx, never poisonous; as wheat, Indian corn, foxtail grass, sedge grass, oats, &c.

Plants whose Stamens *stand on the calyx*, never poisonous; as currant, apple, peach, strawberry, thorn, plum, &c. Some of them contain prus[s]ic acid in considerable quantities, sufficient to render *some parts of them*, at least, poisonous.

Plants with *cruciform* flowers, rarely if ever poisonous. As mustard, cabbage, water-cress, turnip and the like.

Plants with *papilionaceous* flowers, rarely if ever poisonous; as pea, bean, locust-tree, ground nuts, clover.

Plants with *labiate* corols, bearing seeds without pericarps, never poisonous; as catnip, hyssop, mint, motherwort, sage, marjoram.

Plants with compound flowers, rarely poisonous; as Sunflower, dandelion, lettuce, artichoke, burdock.

Plants bearing strobiles are never poisonous, as pines, cedars &c.

Monodelphous, or columniferous plants, are never poisonous; as hollyhock, mallows, geraniums, &c.

Plants with five stamens and one pistil, with a dull-color *lurid* corol, and of a nauseous, sickly smell, always poisonous; as tobacco, thorn apple, henbane, nightshade. The degree of poison is diminished where the flower is brighter colored and the smell is less nauseous; as the Irish potato is less poisonous, though of the same genus as the nightshade.

Umbelliferous plants of the aquatic kind, and of a nauseous scent, are always poisonous; as water-hemlock, cow-parsnip, water-parsley. But if the smell be pleasant, and they grow in dry land, they are not poisonous; as fennel, dill, corriander, Sweet-Sicily, &c.

Plants with labiate corols and seed in capsules, frequently poisonous; as snap dragon, fox-glove.

Plants from which issues a milky juice on being broken, are poisonous, unless they bear compound flowers; as milkweed, dog bane, euphorbium. Lettuce and sow-thistle are milky but they have compound flowers.

Plants have[ing] any appendage to the calyx or corol, and twelve or more stamens, generally poisonous; as touch-me-not, columbine, crow-foot, nasturti[um], monkshood, hellebore.

As a general rule, plants with few stamens, not frequently poisonous, unless they are umbels: but if the number be twelve or more, and the smell nauseous, heavy and sickly, such plants are generally poisonous, and not fit for food or medicinal purposes.

Note. Many plants possess some degree of the narcotic principle, which are still by no means hurtful. As lettuce, Sonchus, dandelion. The roots of some are wholesome, while the herbage is deleterious. As parsnips, potatoes &c. On the other hand, some roots are poisonous while the herbage is used extensively for food. As poke, yellow dock. Plants having a very pungent taste are seldom poisonous; as capsicum, prickly-ash, Indian-turnip, horseradish, onions, ginger.

In our prairies and woodlands, I find quite a number of indiginous plants, bearing the cruciform, papilionaceous and labiate flowers and other indications that they are not hurtful, which by proper culture, might be useful as food and medicine.

On the subject of Texas botany, I may find time to write again.

It was through the entry point of the medicinal uses of plants that Lincecum first began to prepare himself to study botany in a more systematic fashion. He had utilized the standard texts, such as Darby's *Botany of the Southern States* and Chapman's *Flora of the Southern United States*, to learn the language and nomenclature of the discipline.[5] In a letter to W. A. Wilson of Caddo, Texas, October 27, 1861, Lincecum wrote:

I am pleased that you have prepared yourself and have resolved to make a bottanist of yourself—You will find the preparatory steps dry, and of but little interest. But when you succeed in mastering the botanic language there is nothing that can afford, such a mind as you possess, so much real satisfactory enjoyment.

What could afford the lover of nature so much pleasure, as to be able to stroll out into the dark shady forests, at a leisure time, and converse with every tree and shrub, twining vine, tender herb and blade of grass, he may chance to meet with? Talk about "searching nature, up to nature's god," as the dummy superstitionist have it! In the language of the plants, we converse with God himself, and the answers to our questions, are not such as we find in the pasteboard and calf skin records. Put the questions right, and the answers are not only beautifully instructive, but they are profitable, and of a life-sustaining character.

Lincecum's knowledge of botany was rooted in the practical application of plants, not in the scholarly descriptions that academically trained botanists were engaged in. Consider an excerpt from his letter to the editors of the *Southern Cultivator*, September 25, 1860:

Very likely, they may have collected specimens of the most of them, sent them on to some northern institution, where they will be placed in scientific classification with a short description, in Greek and Latin, of the shape of the *panicles, spikes, glumes, paleas &c. &c.*, all of which may be well enough in its place, but of no possible advantage to the practical agriculturist. He requires a discription of its qualities, uses, the kinds of soil upon which it is found to flourish best, and plenty of its seed; and if it is a paying species. He will work out and present the world with it[s] agricultural advantages in short order.

Lincecum became a leading authority on the Texas grasses and was frequently consulted for his expertise in the types of grasses to plant in Texas and the South. On April 11, 1861, he wrote Elias Durand:

I wrote a little article, which was published, advocating the doctrine that the herbs and grasses with which nature intended to feed her cattle with in Texas are already here; and that to be sending off, and paying high prices for the humbug, "morus multicaulis"[6] seeds of distant latitudes, is no manifestation of a good saving, agricultural capacity. The idea took with the Texans and southern people generally, and they have called on me to make the sellection and show them. And as I do nothing this year, but wander, like a digger Indian, hunting the "pismire" [ant], I can do that for them without much trouble. You will now see the reasons why I am desirous of having a corrected nomenclature of the grasses at least. I have already this spring preserved specimens of 25 species of splendid grasses, for hay and winter pasturage.

Lincecum summarized his investigations of Texas grasses in a long letter dated June 1, 1860, to the editor of the *Texas Almanac* as a contribution to the education of Texas farmers and agronomists. This letter appeared in the *Texas Almanac* of 1861.[7] Lincecum wrote:

Now, that all the world "and the rest of mankind" are coming to Texas, it behooves those who intend to remain here to look around them and see what portions of nature's wide-spread bounties can be saved from the destructive tramp of immigration. First, as most essential, I would point the attention of the investigating portions of our community to the analyzation and preservation of the best species of our great variety of superior indigenous meadow grasses; for it requires not the spirit of divination to see that the increasing number of farms, and with them cows, sheep, and other stock, aided by the insinuating action of the destructive plow, will soon put an end to our heretofore boundless fields of luscious pasturage. No country on earth could compare with this, as a stock-raising region, previous to the devastating tract of the incursive plow. It is plain, that our wide-spreading prairie pastures will soon be gone; when we shall be forced to resort to the grass-growing system, or our rich milk and butter and fat cattle will be gone too.

With this subject in view, I have, during the past six or seven years, been examining and experimenting with several species of our native grasses. The result of these experiments clearly demonstrates, to my mind, that any farmer who desires it, may have a first-rate meadow, with but little labor, in the course of two or three years.

Some of our more thoughtful farmers, men whose minds and souls are not wholly engrossed with the all-absorbing "cotton, cotton, first bale of cotton," (hurrying it into market, thereby furnishing the earliest means in their power, in aiding their enemies, the fanatic specu- lators of the North, to trample upon their constitutional rights,) are already beginning to speak of the waning grass, and that it is time we were thinking about sending off to the *North* for the *right* kind of seeds, wherewith to stock our surplus lands with good grasses, before the prairies shall all be plowed up. They say something of experiments they have heard of, as having been made in Texas by some thrifty farmer, with wonderful success, in the cultivation of foreign grasses. One gentleman took me into his garden to see and examine a small experiment he was making with the rescue-grass[8]—said to have come from Georgia—a tolerably good-looking meadow-grass. But the beauty of the experiment was, in the disclosure of the fact, that he had plowed up a pretty fair crop of the same grass (it is indigenous to La Bahia Prairie, if no further) to give room for his costly seed, which he has obtained at considerable trouble and some expense. It was doing finely in the deep mold he had prepared for it, but I could see, close around in the garden, several other species of a far better quality, and which would so have proved themselves had he bestowed upon them simi- lar attention.

It is all nonsense to talk of bringing to this climate the grasses of the more Northern latitudes. The grasses best suited for meadows in Texas are already here, vastly superior for summer and winter grazing and for hay to any that can be brought from other and colder coun- tries. True, they may be grown here to some extent, but never equal to our natives.

Any sufficient quantity of our ordinary black soil, and it makes no difference how closely it may seem to be eaten out, properly in- closed to keep the stock from it two or three years, will show itself to be stocked with twelve or thirteen species of good grass for hay; it will, in fact, be a fine meadow. I have a meadow containing thirty-five acres of that description; it is now ten years since it was inclosed, and, notwithstanding the fact that it never has been plowed, and that we have annually taken off from 30,000 to 50,000 pounds of good hay, it is getting better every year. We are now mowing it, and dry and unfa- vorable as the season has been, a hand can and does mow per day, what will make from 1000 to 1200 pounds of dry hay; and the proof that the hay taken from this meadow is as good as need be, is satis-factorily demonstrated by the greedy manner in which horses, mules, oxen, cows, sheep devour all they can get, keeping them fat through the winter, while during crop-time, the teams that are fed on it keep in as good order as they would on the same quantity of fodder or millet.

Those who can not be satisfied with the kind of meadow I have described above, may, by a little attention at the proper season, pro-cure seed from the inside corners of their own fences, (any where west of the Brazos,) superior in quality for hay and for winter and summer pasture, to any that can be brought and grown *here*, from any other climate. It is not very likely that nature, in the distribution of the seeds of her plants, committed the blunder of sowing any of them in the wrong latitude. The mistake lies in our aptitude to think more of articles of foreign growth than we do of our own. This error is some-times costly.

From the innerlocks of my fences and protected places, I have reclaimed thirteen distinct species of good meadow, or cow-fodder grasses, and there may be more which flourish here, when protected from the cattle, as fine as could be desired; I will also venture to say, better than any grass that may be introduced from other regions un-der similar culture. For all foreign grasses that land must be carefully prepared. Do as much for our indigenous species, and the difference in favor of the native will be very conspicuous.

Four species of the grasses I have selected are biennials, nine of them perennials; five of them are good winter grasses—one superior, for winter pasture, to any yet discovered. Three of them are of the highest class for hay, seven others make very good hay. I send here-with cured specimens of each kind. It will be an easy matter for you to judge of their different qualities and value. Now they are fresh, not having been damaged by packing and consequent sweating, the dif-ference in them is very perceptible. After you get them in hand ex-pose them, separately, a few minutes in the open air, so as to let the compound odor escape, when, by smelling of the different speci-mens, you may readily judge of their various qualities.

I have not attempted to *place* the undescribed species, consequently you are only furnished with their common names.

For a detailed listing of the thirteen grasses, see appendix 3.

Additionally, Lincecum was very interested in sharing his knowl-edge about how to start the grasses, where to locate seed, and what were the best growing conditions. For instance, in a letter to the editor of the *Southern Cultivator*, September 25, 1860, Lincecum wrote:

All that is required of the Texas farmer, anywhere west of the Brazos, is for him to enclose 20 or 30 acres of good prairie soil; keep the stock from it, and in the course of 2 or 3 years, he will have a good meadow, carpeted with a full crop of the original—grasses, such as our prairies were clothed with previous to the incursive track of the destructive plow, capable of producing crops of better hay than can be obtained from any species of grass brought from northern latitudes. It is customary with our farmers, when they obtain, at a high price, some species of foreign grass seeds, for them to prepare the ground in which they intend to sow it, in the best manner: manuring, plowing, pulverizing, raking &c. in gardenlike style. Do as much for some of our indiginous species, and the difference in favor of the native will be so obvious that, in the mind of any reasonable lover of his country, it would put to shame his predilections for foreign things.

With Lincecum's knowledge of medicinal plants and his belief that they grew naturally in the locale where individuals needing them resided, it is no surprise that he proposed that Texas grasses were the best for Texas soil.

Lincecum was consulted not only by Texas farmers but also by individuals from the southeastern states. In June, 1861, he corresponded with I. A. Mitchell of Leasville (probably Leesville), South Carolina, and W. A. Dunn of Winfield, Georgia, describing the planting and cultivation of grasses. For instance, to Mitchell (June 13, 1861) he suggested:

> Anyone, having sufficient grounds for summer and winter pasture and for ample hay meadows, once properly set with the three species—*Agrostis, Stipa* and *Cynodon*[9]—would have but litle reason to pester himself about other grasses for many years; they are all strong rooted, perennial grasses, and bear grazing well. The *Stipa* and *Agrostis* from the first of October through the winter, and on to the first of June. The *Cynodon*, throughout the summer and autumn. Another valuable trait in the character of these three species is that where a man possesses sufficient control over his averice, they pro[pa]gate themselves forever, after the ground has been only fully stocked.

In the letter to Mitchell, Lincecum suggested that in 1859 he had recognized a new grass, *Cynodon dactylon*, a double-bladed panicum from Bermuda. Lincecum wrote:

> There can be no doubt of the fact that this grass made its first appearance in this county or state in 1859. I have been investigating the grasses of Texas during the last twelve years and should not have failed to find it, if it had been here. Neither has any one else, that I have spoken to on the subject, seen it previous to 1859. It may be true that it is a

native of Bermuda. Could it not also be a native of our southern coasts? If it has come here on its own accord from any country, is it not as good as a native? That it is travelling from the west, is proven from the fact that it is only found near the old San Antonio road, and that it is now found but a few miles east of this place. It is a most beautiful grass and is devoured by all the grass eating animals. Its roots are perennial in this latitude.

To W. A. Dunn (June 19, 1861) Lincecum wrote:

I shall enclose herewith six species of our good natives. The seeds of these grasses may be kept on hand until the fall rains have put the ground in order for the plow, when they should be carefully put into the soil. Say, in October.

The safest chance to ensure the propegation and secure the largest crops of seed from the *Stipa setigera*, *Bromus corinatus*, and the double-bladed panicum is to sow them on the borders of some of your walks. They make beautiful border grass, and being where you will be sure to see them often, you can be certain to preserve the seed as fast as they mature. The stipa has no superior as a winter grass, except the agrostis; and that has been disputed by Texans. The *Bromus corinatus* produces fine hay, heavy crops of grass, and is a pretty fair grass in a mixed pasture for early spring grazing. In this latitude it is a good winter grass, but will not, perhaps, stand your winters. The double-bladed panicum is the best summer pasturage we have. Horses prefer it to all other grasses and will, where it is not plenty, gnaw it into the very ground. It does not grow very large; propagates itself by rooting at the joints and by heavy crops of seed. It is by far, in my view, the most beautiful among the grasses. Its roots survive through our winters. *Bromus ciliatus* is our upland Wild oats; flourishes well in our stiff post oak lands, having a clay foundation; is fit for the scythe by 10th. of May; grows two to three feet high; and has strong perennial roots. *Phalaris intermedia*, if sowed in time, comes up about the first of December. Through the winter, it looks, tastes, and smells exactly like young wheat, and is perhaps no better for grazing. It however produces fine hay and quantities of seed that will delight your canary birds. The *Trifolium Texana* is our little white clover. It spreads itself over all our timbered lands; by the middle of February it begins to unroll its green carpet, and by the first of March, hogs and all the stock are grazing on it. By the tenth of May it is all dead, and everything fat. Our hogs, the past spring, were *big* pork on it. It does not rise exceeding six inches, but is mighty in quantity, bearing much seed.[10]

To the editors of the *Southern Cultivator*, October 15, 1861, Lincecum wrote of another grass that had appeared a few years before in the Austin area. He described the growth of *Panicum gibbum* (*P. obtusum*; Austin grass):[11]

> The *P. Gibbum* was first noticed, several years ago, on the cultivated bottom lands of the Colorado, in the vicinity of Austin. It made its appearance in the fields after crops were laid by, and the farmers in that region were begining to fear that it would turn out to be a troublesome visitor. But about the first of August last year, when, from the effects of the protracted drouth, the whole country was entirely nude of every species of grass, stock of all kinds were subsisting on a scanty supply of bushes and brush; and when there was no hope amongst the farmers that there was any chance to prevent their numerous flocks of cattle from actual starvation, the rain set in—8th. of August. In four or five days this wonderful grass was seen springing up over all their sun scorched fields, thick as wheat. In twelve days, cows, horses, sheep and hogs, could begin to fill themselves on it; and by the middle of october, stock of all kinds looked as well as usual; the people were making hay of the God-send grass, and there was such quantities of it that the farmers could afford to take it to town and sell it at from 6 to 8 dollars for the largest kind of wagon loads.

Lincecum's attitude toward the Northern scientific establishment can be easily seen as he continued in the letter: "I sent specimens of this grass to the Academy of Natural Sciences at Philadelphia. They say it is a new species and place it as an offshoot of *Panicum obtusum*. Well, it is a regular *Panicum*, and for convenience, until I am forced from the position, I shall give it as a specific name, *Gibbum*. I have as good a right to name a plant as any yankee; especially southern plants; and so in my nomenclature, for its scientific name, it stands as *Panicum Gibbum*, and for its common name Austin grass."

Lincecum was also a proponent of conservation of Texas grasses. To B. Shropshire of LaGrange, Texas, April 7, 1861, he wrote: "Therefore, I am pleased to see the spirit for preserving our *native grasses* be so thoroughly indicated and so wide spread in the southern states. A meadow, for summer and winter grazing and for hay, has heretofore been but little thought of by the southerner. They are however waking up to the subject now, and ere long they will demonstrate that our soil and climate can produce as good grass as any country on the globe."

In 1868 Lincecum was still arguing his case about the native grasses of Texas. In "The Indigenous Texian Grasses," he wrote for the *Texas Almanac*:[12]

Eight years ago I wrote an article concerning the grasses of Texas, that was extensively circulated in the periodicals and newspapers of the South; but because, as I suppose, it was done up in the language of the botanics, it was but little read. I will write again—not from the sympathy I entertain for the careless, lazy owners, but for the poor cows and horses that will be forced to bushwhack for a living here, long after I have passed away to the "good hunting ground;" and I will endeavor to perform the writing without the use of a single technical. . . .

Lincecum was also very interested in Texas grapes, and *Vitis lincecumii*, a bush grape, bears his name.[13] In a letter to S. B. Buckley, September 25, 1866, Lincecum described three species of post oak grapes he sent to Elias Durand:

I sent off, yesterday morning, by mail half an ounce of dry post oak grapes to Durand. They were nicely dried, pretty good raisins. Durand wants them, to send for seed grapes, to france. There are three very distinct varieties called "Post oak grape" by the people. The first is the bush grape, or *V. Lincecumii*. It delights in dead beds of sand, and its fruit falls off as soon as it is ripe. The second is a tall vine, found in sandy post oak lands, running over the tops of the trees bearing a large black grape, which is nearly as harsh as the mustang grape, and it is my opinion that it is a hibrid betwixt the mustang and the *V. Lincecumii*. The third is also a tall running vine, found in sandy post oak lands, bears a pretty large edible grape, which remains on the vines until it is eaten by the birds, or otherways destroyed and is I think a hibrid betwixt the harsh vine grape and the *V. Lincecumii*.

In January, 1866, a brief extract on Texas grapes from a letter Lincecum wrote to Durand was read before the Academy of Natural Sciences.[14] Lincecum also corresponded with Dr. George Engelmann on their mutual interest in grapes. On October 20, 1867, he wrote: "You ask if I have studied the grapes of Texas. I may reply, that I have, first and last, collected every species of indigenous grape belonging to Texas and sent them to Durand. I do not think that there is a single species left. On the Guadaloupe river I found a good many plants that do not flourish east of the Colorado, which has, previously to last spring, constituted my botanic range in Texas."

Additionally he corresponded with T. E. Johnson of Brenham, Texas, on January 26, 1859, concerning the medicinal use of mustang grape wine. Lincecum wrote, "I have prescribed it in quite a number of cases of protracted intermittents—in the stage of exhaustion from severe typhus and some nondiscript cases of low debility, with decided advantage, which generally begins to develope its good effects in a very short time. My experience with the wine produced from the Mustang grape, in the treatment of disease, runs through a course of ten years."

Lincecum's interest in botany was wide ranging. In a letter to Professor Alpheus S. Packard of the Essex Institute, Salem, Massachusetts, February 18, 1868, he responded to an article published by the noted Harvard botanist Dr. Asa Gray in the *American Naturalist* concerning *Tillandsia usneoides* (Spanish moss). The letter is revealing of Lincecum's independent investigation:

> I noticed an article by A. Gray, headed "Botanical Notes and questions["].—It occurs at page 673, of the "Naturalist" [Vol. 1 (1867)].
>
> To his question, "Is *Tillandsia usneoides*, the black or 'Long Moss' of the southern states, strictly an epiphyte, or is [it] some sort of a parasite?" It is clearly an epiphyte. I have gathered it to fodder the cattle with myself, hundreds of times, and after reading Mr. Gray's article on the subject tonight, to make all sure about it, I lighted up my lantern and, going out to the forest, broke off two branches, equally loaded with the *Tillandsia*; one from a dead Post Oake—*Quercus obtusiloba*[15]— the other from a live one. I carried both into my room, where I examined them before a good light. The result of the investigation was that on the dead tree and live one both it had taken root in the dry and decaying bark. In some places on the live tree, its roots had gone through the dry bark and did seem to be penetrating the moist green bark. On the dead tree, the roots, after passing through the old, dry outside bark, had actually penetrated the dry decaying wood. In both cases the moss is equally flourishing; I can see no advantage to the moss in the rooting process, but to enable it to cling to the body of the tree and to the limbs that are too large for its filiform stem to wrap around. On the small twigs of the live trees, there are no sign of roots to be seen; but you will see its dead, hairlike, black stem wound around them, and they are sufficiently strong to suspend long, beautiful festoons of the living *Tillandsia* for years. I know of many trees that have been dead five or six years, and the moss is as heavy and seems to be a[s] flourishing as that on any of the live trees, if not more so. As to the statement that it will wither and die when the tree upon which [it] hangs is cut down, I can only say it is not so in this country. I have seen many trees cut down, and the moss on the limbs that remained elevated above the reach of the cattle would be covered with it, and it would flourish the same as it did on the standing green trees. The *Tillandsia* may be taken from one tree and thrown on the branches of another and if it is tied on so as to prevent the wind from blowing it down, it will continue to grow as luxuriently as it did previ-

ously to its removal. I have seen large quantities heaped up for it to heat and rot the bark from it—for that is the way they prepare it for Market—the outer coating of these moss heaps could remain alive and continue to grow for months. Occasionally I have observed trees that were so heavily loaded with it that they seem to die from suffication, but the moss continued to grow on. If the *Tillandsia usneoides* contains "*Silica, alumina*, lime potash, sulphates &c.," it does not obtain it from the trees upon which it hangs. The live twigs of all the year old moss are hanging to dead threads 3 or 4 inches long, specimens I enclose herewith. The part that has the thread tied on it is the part that lay upon the limbs and you see it is dead. I enclose also a specimen from the dead tree, which has rooted itself into the but[t] of a dead, oak branch; the moss is dead at the connection.

Like his early mentor, Erasmus Darwin, Lincecum could wax poetic and use his imagination in studying the plant world and its origins. For instance, consider this visionary account of the planting of one particular black oak he found in an unusual location.

No. 160 Black Oak—*Quercus*
Sunday, 28th October 1860

Until today, I had not seen this [species of] oak South of the Trinity river. I found it growing in a dense thicket on a little unfrequented point, overhanging the ravine which runs down from, and not exceeding half a mile below, Long Point, Texas. It is a double tree, rising from the same root, which seems to have been the stump of a former tree. The trunks of the two trees, separate near the ground, are of equal size—30 inches in circumference and 70 feet high. They have produced no fruit the present year; consequently I can send you specimens of leaves only.

There are no other *Quercus Tinctoria*[16] to be found in a hundred miles of this place. With me the question arises as to how came the acorn, from which these trees have their origin, on that little woodland point? A crow would not carry it so far. A squirrel couldn't; the mallard duck, who is a great acorn eater, does not feed on the acorns of the black oak. But the wild Pigeon does; and now, as to the how he happened to leave the acorn there? Well, for the purpose of ascertaining the obscure facts, in regard to the history of the origin of these, now thrifty, oaks, I work myself up to the Clairvoyant condition, and while in this elevated state, fly swiftly back through the ages gone, until I reach the fruitful year A.D. 1701. 'Tis November. Scenery, a woodland, bordering on a far reaching prairie to the South. Upon this gently undulating sea of grass is lazily feeding an immense herd of fat Buffalo. It is a sweet, bright day, with a brisk, but balmy south wind, sweeping over the rich, wavy grass. Nature seems to smile as she feeds the Buffalo and peace and plenty prevails over the widespread plains. Peace! A state of quietude, spoken of in the homilies on the subject of our imaginary heaven. But there is no peace! But there are many empty stomachs. It is the arm of hunger that hurls the shafts of mischief and death, and this great warring world is but one vast slaughter house.

Anon is seen approaching from the East, a flock of wild pigeons; whirling and dashing and disorderly crowding; the affrighted flock wildly sweep on, in their erratic course, through the sounding air. Close in their wake, in hot pursuit, with half expanded pinions comes the keen winged Falcon. Already he has marked his prey. And now with increasing speed, he wings his unerring course; the bird perceives his vile intent and deems it best to quit the gang; quick as thought the hawk flits onwards, and pressing nearer to his hapless victim, now whirling downwards, as the last resort. The hawk and bird together strike on the woodland point, where now stands the black oaks. Mercilessly, the falcon tears the inoffensive bird, which soon yields up its harmless life. The throbbing flesh is greedily devoured; with voracious gorge the feast goes on till naught is left save the well filled crop. Upon this he clamps his bloodstained beak, and rudely shaking, and slapping it around, 'gainst the brier and bush and brush, slings out the unwanted contents, which the slaughtered pigeon had dilligently picked up, in a far off country—Among other things, that pigeon's crop contained Black Oak acorns.

In a similar vein, Lincecum challenged young W. A. Wilson (in a letter dated October 27, 1861):

You ask the rugged, unsightly oak, what can you do? The oak replies, I can afford food for hogs, squirrels, rats, and numerous birds, and that is not all [I] can do. Question me with impliments of tempered steel. I answer you in boards and timbers for the builders use; in posts, stakes, and rails to enclose your farms and make your gates; in timbers for your ploughs, wagons, and carriages; I can furnish ample fuel to warm your houses during winter, and ships to float the world's commerce, and [withstand] her thunders upon the briny wave. And that's not all the genus *Quercus* can do. For a thorough knowledge of my uses: [I] will tan your leather; furnish you with ink, black dye, and many other smaller things, besides entering in to your catalogue of medicines, where I am famous as an astringent, giving strength to the relaxed fibre and mucous surfaces; and to the desponding, feeble female, restore to the productive organs strength and vigor, constitutional health and vivacity. How would you do without me?

In contrast to the poetic, Lincecum also wrote some very detailed accounts of practical uses for plants. For example, here are his directions for how to dye cloth, written in a letter to Mary H. Taylor (March 26, 1863), of Navarro County, Texas:

Most respectfully, I reply to your application on the subject of the cheap coloring material of Texas, and the processes as far as my experience goes. The gentleman who reported my experiments to you was incorrect in respect to the alum; I cannot produce that very useful mordant on terms that would pay; nor have I as yet been able to produce the Turky red or indigo blue from any plant I have tried so far.

Coperas, or acetite of iron, which, in coloring cotton is superior to the sulphat of iron (coperas) because it never damages the goods — and it is fully as good in coloring wool — may be had to any desirable quantity by the following process. Throw into some vessel that will hold it, all the old waste iron about your place: scraps of tin, old hoop iron, old stove pipe, any iron scrap. Then fill up the vessel with weak vinegar, or sour beer; set it in some safe place. It soon becomes a fountain, from whence you may draw at any time, and the longer it stands the better and richer it gets. It should be kept filled up all the time with the weak vinegar or sour beer, and the liquor that is left, after dipping your goods, should be poured back into your coperas fountain. It is not proper to put this coperas liquor in with your bark, but saturate your cloth by steeping it in the liquor (I'll call it vinegar of iron) — for black dye, two days, stirring and wringing often. Then place the goods in your coloring decoction and turn it, lifting it often, until you have the desired shade. For lighter shades, keep it in the vinegar of iron a shorter time.

The vinegar of iron is applicable in all cases where you would use the coperas, and it is equal to it in dying wool, superior for cotton or silk, and much more convenient. The ladies are all familiar with the process of coloring black, brown, and the various shades of these two colors — I need not say more on that subject.

To color drab, and light reds, the ley [lye] of some strong ashes is the proper mordant, and it is applied after the goods have been boiled sufficiently in the coloring matter. The ladies all know how that is done. The brightest red I have produce[d] on cotton was done with the red elm bark, the wild peach bark next; the goods being washed in ley after thoroughly boiling them in a strong decoction of the bark.

The bright yellows cannot be produced without alum and acetite of lead. The alum you will have to procure from the druggists; the acetite of lead you can make at home. Put a pound or so of lead into a glass jar (shot or finely chopped up lead)—an old pickle bottle will answer—fill it up with good Vinegar; let it stand in a warm place five [or] six days, shaking it often when it will be fit for use. The process for dyeing yellow, a specimen of which you will find herewith inclosed, is not difficult, but it requires that you be exact as to quantity in the proportions of your mordants.—for wool: let the wool be boiled for an hour, or longer, with about one sixth of its weight of alum, dissolved in sufficient water to keep the goods wet while boiling. Then, right from the alum bath, without rinsing plunge it into a strong preperation of black oak *(Quercus tin[c]toria)* bark. That is, pretty finely bruised black oak bark in quantity half the weight of your goods, put into sufficient warm water to cover all fully; into this preparation the wool is to be placed, and turned and pressed in the liquor, at the same time raising the heat gradually, to the boiling point, which must be continued until the desired degree of color is brought out. Then sprinkle into the liquor a quantity equal to one hundredth part of the weight of the cloth, of clean, finely powdered chalk, and continue the stirring and turning and boiling 8 or 10 minutes longer. By this time, a pretty, bright, lively yellow will be produced. Wring it out and dry it, after which wash it with soap. Wring it well and spread it out to dry again.

The process for dyeing cotton or linnen yellow is a little different. Make a saturated solution of alum, and to three parts of this solution add one part of the acetite of lead. This solution must be heated a little above blood heat and kept at that temperature; the cloth or hanks should be soaked in it two hours, then wrung out and dried. The soaking may be repeated, and the cloth or hanks again dried as before. It is then to be barely wetted with lime water and dried again. The soaking in the alum and vinigar of lead may be repeated, and if the shade of yellow is required to be very bright and durable, the alternate wetting in the lime water, and soaking in the alum and lead may be repeated three or four times.—Then the dying liquor is prepared by putting twice the weight of your goods, of finely bruised black oak bark, tied up in a bag, into a sufficient quantity of cold river or rain water. I mean to say any soft water. Into this water with the bag of bark, the cloth or hanks are to [be] put and turned in it for an hour, while its temperature is gradually to be raised till you can't hold your hand in it. It is then to be raised to the boiling heat, which must be continued but 3 or 4 minutes; if boiled to[o] long, it will aquire a shade of brown.

The pacan *(Carya olivaeformis)* bark, treated in the same manner, as directed for the black oak bark, will produce equally fine yellow colors. The bark of the spanish buckeye *(Daubentia)* by boiling the goods with the bark in water two hours, and then washing in ley produces a very pretty drab color. Red Elm (*Ulmus ameracana*) treated in the same way, colors cotton the redest of anything I have yet experimented with; next best to it, is the bark of the wild peach (*Cerassus cardinaensis*). This article colors wool a most beautiful bright brown, almost a red color. It will also color cotton from a very fair nankin, through many shades to a good arnetto; this dye is treated with lye also. The best black dye material is the heart wood of the mezquit, (*Algarobia glandulosa*).[17] Chip the wood very fine and treat it with the vinegar of iron, just as you would logwood. The mezquit wood is equal, perhaps superior for coloring dead black, to the logwood itself. We have a small weed in our prairies, the perenial roots of which are red, capable of producing many shades of color on cotton goods, from a pretty bright red, through the various phazes of brown, to a dead black. I have experimented with many other barks and roots, many of which produce bright and useful colors for stripes &c. But the prettiest, the cheapest, and most abundant of all the coloring material for coloring wool is found in the poor land green moss, which so profusely drapes the scrubby post oaks, elm, and other trees growing on poor glades and stoney knobs, almost in every climate. It can be had in any desireable quantity. . . .

The process for dyeing with the green moss is so simple. . . . An iron pot will do. Fill it with clean water, put in a layer of moss, then a portion of your cloth, or hanks, or wool; then another layer of moss; then a layer of your goods, and so on, alternating the moss and the goods until your pot is nearly full. Now put your fire under it, and after you have raised it to a boiling heat, commence stirring and turning and pressing it down, and continue boiling for the space of an hour and a half. Then wring it out, wash it, and when it is dry, you will have a very pretty and very durable bright, but rather light brown. If you are not satisfied with it, car[r]y it through the same manipulations with a new lot of moss; and the process may be repeated even four times. It will receive a deeper shade every time. I admire the deepest shade above any color. Inclosed you will find a lock of wool, dyed the third time.

Who but Gideon Lincecum would have taken such pains to give this woman a *course* in dyeing?

A practical application of a different sort involved syrup and wine making. To George H. Vaughan, of Prairie Lea, Texas, Lincecum wrote (June 11, 1860):

My experience in the production of Syrup and mustang wine is not very extensive. I have however, succeeded in producing both, of pretty good quality; at least travellers say so.

My method of manufacturing the Syrup is quite simple. In the first place, let the cane be ripe, but not too ripe. Press out the juice, straining through a common wire or hair sifter as it runs from the mill; it must not be suffered to remain long after the grinding, before it is put into the boilers. It is the best way not to have more juice ahead of the kettles than you can boil down the same day. After the boiling has commenced, it must be constantly attended to—stiring and skiming—and it must not be suffered to boil down as low as the line where the fire strikes the boiler. To prevent this it must be often filled up with fresh juice. The object is to prevent burning the syrup, which will ruin it by making it black and bitter; hence the necessity for keeping it constantly stirred, with the juice all the time above the fire line.

All the foam must be taken off with your skimmer, from time to time as it rises, and if necessary it must be strained before it gets too thick. To divest the juice of all impurities, which is easiest done while it is boiling, keep it from burning, which it will very soon do if it is not constantly stirred and the boilers kept full enough. And to boil it down until it is a thick, heavy, ropy syrup, is all the skill required in the process. A common water bucket with a long handle inserted diagonally through it is a good & convenient vessel to lift and stir it with. The skimmer may be made of copper, purforated with small holes. Tin will answer very well. There has been a great deal said about the use of lime in the process. I think it is altogether useless for making Syrup. The sweetest and the brightest we have was produced without a particle of lime or any other chemical. To cleanse and boil it down sufficiently without burning, is the secret of success; and this cannot be done without unceasing, vigilent attention.

To make wine from the mustang grape is still more simple. Let the grapes be ripe. Gather them clear of leaves. Put them into your vat, barrel, or whatever vessel you have, mash them with a wooden pestle, adding grapes and mashing until your vessel is full to 3 or 4 inches of the top. Tie over this top of the tub a blanket or other cloth closely, to keep the gnats out; place it in the shade, and let it remain quiet 25 days. Then draw off from near the bottom all that will run without pressure; strain it through two blankets into a clean barrel. Put a spile [spigot] in the barrel, set it on skids in a cool place, where it must remain perfectly quiet 30 days. Then carefully draw it off without agitation until the sediments begins to appear. Strain it into a tub or any open top vessel; add a pound of good dry sugar, and half an

ounce of finely pulverized chalk for every gallon. Now agitate it violently with a paddle or dasher for some time—until the effervescence
has entirely subsided. At this stage of the process, if you like, you may
add half pint of good spirits to each gallon. Set it away under close
cover 7 days. Then strain it of[f], bottle and cork it close, and keep it in
a cool place. Clean jugs and demijohns will do as well as bottles, but
they must be full and kept cool. If you gather enough to make it
necessary to use barrels, they must also be full and kept cool.

Determined not to shortchange Mr. Vaughan, Lincecum continued his
instructions: "After you have drawn the wine from the pummis in your
first vats, you may add to the pummis a bucket of water for each bushel
of pummis, press them down level, let them stand under cover ten or
twelve days; then draw if off and barrel it away for vinegar. But the
clean strained juice of the sugar millet, barreled away makes the best
vinegar—half and half of cane and grape juices, will ferment in 48 hours,
producing a very pleasant drink, or table beverage."

Why was Gideon so interested in botany, even beyond its applications to medicine and agriculture? At the age of sixty-eight, he wrote
to W. A. Wilson, of Caddo, Texas (October 27, 1861): "An acquaintance
with the science of botany generates a taste for other branches of the
natural sciences. All of which are calculated to soothe and amuse the
period [of] dull old age."

Chapter 3

Gideon and the Texas Arthropods

To know all about every type of the insect world, down to the lowest microscopic animal scale, demands from the investigating mind, serious, earnest study. Man, having been progressively developed from an animal condition, cannot know himself until he is acquainted with the inferior types of the living and moving world: his kindred forms.

Gideon Lincecum

Given Lincecum's interest in Texas botany and his strong sense of kinship with all living things, it was almost inevitable that his attention would be drawn to Texas grasshoppers.[1] His comments on this subject in a letter to his son-in-law, George Durham, in Austin (May 5, 1858), indicate that he was just beginning to turn his attention to entomology:

> Your quite scientific letter on the subject of grasshoppers and their destiny reached me today at noon. If I understand you, they have already eat up one crop for the farmers of your vicinity. You say, the people are planting the second crop, which signifies as I suppose, that the grasshoppers distroyed the first. I wish we could gat [sic] a shower of them down this way, our prairies positively do not produce a sufficient quantity to afford us a reasonable supply of fish bate. But I will not complain, as possibly we may be favored with a full share of them in their next periodical visit to our land of grass and flowers.
>
> You ask for my ideas on the grasshopper subject, their determination &c. Well, with your species of the genus, I am not familiar, to an

extent that would enable me to say much that would be reliable. Two years ago, last november, I saw clouds of them passing coastwise for three days. Some of them descended to the earth of evenings to spend the night with us, but as soon as they felt the warmth of the morning sun, they took wing again and passed on, in their journey southwardly. The wind all the time blew gently from the north. I caught a good many of them, and weighed and measured them, and that is about the amount of my knowledge of the distructive little pests.

I do not think however, that they will continue long. Did they ever, since you resided in Texas, make their appearance before? I mean, as far East, as the vicinity of Austin?

There are many species belonging to the genus Locusta,[2] any of which would be equally distructive to vegetation, should they become so prolific and make their appearance in such countless multitudes as the species under consideration are now doing. For aught I know, they may—I mean, the different species—take it by turns, and so for a series of years, it may be our lot to submit to the feeding of a race of them until we pass through all the species of the hopping genus. But I have no idea that such will be the case. I am of opinion, that they, like the cicada,[3] have but two or three prolific species; that they make their appearance periodically, at long intervals; and that the present season will terminate their visit to our state, that is, to the middle and eastern portions of it. I think their habit is, as soon as they get wings, to fly away, and the object is to deposite their eggs in other districts of country, not only for the purpose of extending the race, but also, to ensure food for the coming generations. But we may not fear. Nature has ample checks for all redundences. Had she not, these little hopping insects alone, in the brief space of seven years, would depopulate the earth, and like Sampson, die among the ruins they had themselves produced. But overtaken in their distructive course, by fierce Ichneumen, Libellula, "epizooty" or some other plague;[4] or, all together, his destiny is soon to disappear and be seen no more for a *season*, perhaps an indefinite period. We do not live long enough to make observations on a suficient number of grasshopper epochs, to form any reliable conclusions as to the intervals between their visits or the period of their stay, when they do us the honor. All these things however, are ultimates from the ceaseless workings of a perfect law, and are the subjects of regular periodic returns, be they long or short.

After all you see that I have but little knowledge on the subject of grasshoppers, and you must be lenient and forbear forcing me to submit to too close a scrutiny, or you will make the discovery that my actual knowledge is but a small domain, particularly in the bug kingdom.

Writing on the same subject nine years later, Lincecum showed he was by then much more knowledgeable on the subject of grasshoppers. Writing to Spencer Baird of the Smithsonian Institution on July 8, 1867, he reported:

The same cause that prevented me from procuring birds eggs operated unfavorably on the insect world; the grasshoppers that had devoured the wheat crop extensively were exterminated by the severe norther that occurred on the night of 13th March. The first days of March were warm and spring like, and the grasshoppers had hatched out in countless millions, particularly in the wheat region. I saw their sign in Bell, Corryell & Burnet Counties—They seemed to have gathered, somehow [or] other, all to one side of the field, and marching into the growing wheat, consumed it utterly as they went. I saw some wheat fields that had been half destroyed, other fields not so much; and the distruction was so complete that not a speck was to be seen on the ground over which they had passed, and the line of their progress to the point where the norther struck them was distinctly seen, as far as you could decerne the field, and sometimes the line of demarkation was pretty straight. Grasshoppers are pretty scarce thus far this season.

My collection of insects is not as large as I expected. Coleoptera,[5] is pretty fair, and from the indication with the new crop of grasshoppers, I shall be able to make full collections of them. Not being able, on account of its high price, to afford alcohol to preserve them in, I am trying to preserve my specimens in $3 whiskey and alum;[6] consequently I fear I shall lose most of my specimens, particularly the grasshoppers and beetles. This item operates as a heavy check on my collections, and there is no way to get around it.

Writing five months later to Professor Joseph Leidy at the University of Pennsylvania, Lincecum could report that there was no longer any shortage of grasshoppers in Texas (November 13, 1867):

The grasshoppers were on the wing today in countless millions. They were coming from the N. W.—We had thousands here already, that had dropped down seemingly from the heavens three or four days ago. They all rose up today, it being warm, and went off south. Others coming from N. W. were seen, filling the upper stratum of the atmosphere. They were very high and in such numbers, that they appeared to float through the suns Aura in masses, looking like *star-dust*—nebulae. Their glittering glass-like wings, seemed to burn on the very disc of the sun. Towards 3 P. M. they began to discend, and by 4 P. M. we had ten times as many on the ground, all new comers, as had left us in the forenoon.—I could see how it happens that showers of grasshopper[s] occasionally occur at sea.—They are boring holes in the ground and depositing *bushels* of egg[s]. Every square foot over many miles, contains from 10 to 20 grasshoppers, and travellers say that they are comparatively few here.

Five days later he treated Spencer Baird with a much more grandiose description of the great grasshopper swarms and depredations (November 18, 1867):

This is one of our Locust years. I have collected a sufficient number of them and put them in the bottles with the other grasshoppers. So that you may make no mistake in regard to the particular species we have now with us, I have filled two small vials with them and dropped them into the grasshopper bottles. After being placed in the alcohol, they turn red. The excrement of the chickens that feed on these grasshopper[s], is very red, and I have no doubt of its being a good dye stuff. No other grasshopper turns so red when you drop them into spirits, as this peculiar, long winged, red legged species.[7] They have been depositing their eggs in this vicinity about five days, and they have bored up the surface of the earth at such a rate that the whole country has the appearance of having been stirred up by freezing and afterwards been dried in the sun. The eggs are deposited in little encrusted cases of earth which is glued together by a mucilaginous secretion, discharged at the moment of laying the eggs. These black earthy egg cases are 1/8 of an inch in diameter, 7/8 of an inch in length, containing 16 eggs very nicely laid cross ways in the case, the length of the egg giving the case its diameter. And they are deposited in such quantities, that a brisk hand could collect a bushel of them in less time than a day. This is a frightful account, but is certainly true. There are doubtless many bushels of their eggs deposited in a few miles around.

When I was far out N. W. last spring, I saw these grasshoppers in their infantile state; they were but just hatched, and they were so numerous, that when walking along amongst them, I could distinctly feel the weight of the drift of them on my feet, as they were trying to hop out of my way. This occurred when the whole vegetable world was in full foliage, and having plenty to eat, they did but little damage to the crops. This was considered a late crop of grasshoppers. There had been an early crop of them, hatched out about the first days of March, and marching in long extended array, went into the wheat fields (the eggs that had been deposited in the fields had been turned up by the plough and had perished during the winter) and sweeping the young wheat as they went, had destroyed, in many places, half the fields by the 12th of March when there came a severe norther and put a stop to their devastating course. They were not yet larger that a house fly, and yet their numbers were so great that the masses of the dead ones produced a horrible stench. The late ones discribed above however, grew up to manhood without attracting much attention until they got on the wing, when they soon became the talk and the wonder of everybody.

The first I saw were high up in the blaze of the sun, perhaps ten to fifteen thousand feet, yet their glittering wings could be plainly seen as they whirled on their quivering course through the bright halo around the sun.

About 3 P. M. on the day I first discovered the grasshoppers floating in the sunlight, there came a light cool breeze from the north, and in a few minutes, millions of the grasshoppers were whirling downwards to the ground. In less than an hour, the whole earth was literally covered with them. During the afternoon, they refreshed themselves on the green grass, weeds, goard vines, okra and jerusalem oak.[8] At night they took lodging on the bushes, weeds and fences. Early in the morning they commenced their breakfast, racing on, until after ten o'clock, when they took wing again, going South, the ground soon being entirely clear of them. On looking up to see how high they went, I could perceive, in the sun's aura, millions of them coming from the N. W. and so high that their glassy wings seem to burn in the very face of the great luminary. And they were crowded in such vast multitudes that away beyond those that could be individualized, were drifting white cloudes of them, resembling the much talked of star dust, or nebula. Towards 3 P. M. they began pitching down as they did the day before, but these were not our grasshoppers, that had rose up in the forenoon. They were forever gone. But we had a new set of as good looking locusts, and at least ten times their number. In the afternoon of this day, they began to couple, and I noticed that some of the females were boring holes in the ground. They bore the hole with their tails.

These passed away south in the forenoon of the next day, to be replaced, with an equal or greater number in the afternoon, and so they continued to come and go, each day up to the date of the writing hereof. On the third day after their arrival here, they commenced depositing their eggs in good earnest. They began at first, to bore their holes and deposite their eggs in the pavements of the agricultural ants and other places of nude grass. But soon, there were countless millions of them engaged in making their deposits, and now the entire surface of the earth for hundreds of miles is broken up and so full of their egg cases that, scratch[ing] a plane an inch long and half an inch deep, will rake out two or three of them. Our farmers are all frightined at their prospect for a crop next spring. A hard winter will relieve the farmer of his fears.

They bore the hole for their eggs with a small bony scoupe which is attached to the lower division of the vent. I saw yesterday thousands of them, thickley set upon the ground, with their abdomen bent down at a right angle, boring and working it into the ground. It is a slow motion, but they finally succed in penetrating the ground seven eights of an inch, when they soon deposit their eggs.

And now as I have sufficiently bored you about my boring grasshoppers, I will close by wishing you a long, prosperous life.

If for Lincecum grasshoppers were a plague as well as a source of wonderment, honey bees[9] were to him a source of great joy as well as an economic blessing. On April 11, 1861, he wrote to Elias Durand:

> Now the honey bees are swarming and about to settle on a large poast oak. There will be another climbing job for me. Did you ever work with the bees? I have as many hives as can do well at this point, and thousands of them are acquainted with me and will play with me, as familiarly as my cat will. Their mode of play with me seems to be an acquired habit. I never see them act in that way with each other. When I put my hand amongst them at the door of the hive, some of the guards will approach my fingers, and striking them an overhanded blow with the front edge of his wing, whirl around and run off into the box, as if he expected me to follow, and if I push my fingers after, so as to crowd him a little, he will stop and turning up his tail, standing almost upon his head, push out his sting about half its length, as much as to say, "you better take care."

Five years later (June 16, 1866) he commented to Durand:

> You speak of Huber[10] and the Honey bees. I have seen notices and extracts from his *bee* history often, but have not seen his book. Many of the extracts I have read did not discribe the manners and customs of our American Honey bees, which, as I suppose, is attributable to the bees in his country being of a different species. Our bees are susceptable of training and can be taught extensively. I have, at various times, trained the surplus bees that were lounging about the base of hives, to go to work in places outside, and entirely unconnected with the main hive. For instance, I placed a piece of new honeycomb in an old churn, laid it on its side near where the lazy little fellows are idling away their time, and by holding the honey comb among them until it was well covered with bees previous to placing it in the churn, repeating the experiments until they became familiar with me and

my oft repeated manipulation, they went contentedly to work, and soon filled up the *small* churn with extremely delicate combs and colorless honey. I noticed one other peculiarity about this delicate honey comb. There were no deposits of the pollen of the flowers (bee bread) found in it. This I attributed to their knowledge of the fact that it was not a permanent establishment, that they having no queen in the churn, did not intend to feed and raise young bees, and more, that they did not intend to remain long in it, for they left it as soon as the nights became a little cool and went into the main hive. On removing the churn it was found to be full of sealed up honey, and not a single bee to be seen. I have directed the action of the idle portion of the hive in several other occasions, causing them to fill little wooden boxes, and in one instance made them fill a sardine box. I consider them quite intellectual and very capable of receiving instructions from the human genus.

In a long letter to the editors of the *Southern Cultivator*, December 1, 1861, Lincecum offered the readers of the publication considerable advice on bee culture, especially to limit the depredations of the bee moth.[11] "I have eight Bee hives," he said, "which have been hanging on the same racks for the last twelve years. They have been robbed [by Lincecum] of their honey every year; and they are now, vigorous, full swarms, their boxes well *packed* with honey and bee bread for winter supplies. The moth, though very abundant here, has not disturbed them." He went on to give detailed instructions (and even diagrammed) how to make a moth-proof hive and then added what precautions needed to be taken in "robbing" a hive. The latter are especially interesting:

When you rob them, first tie a cloth about the lower end of the box suffering it to bag down considerably below. Then lay it on a table, with the head an inch or two the lowest, so that the dripping honey will run towards you; cut the combs smoothly, and even down to the stick, but no further. Scrape the sides of the box, and wipe the honey out with a cloth; put on the head and hang it on its rack again; untie the cloth, and dropping it beneath on the ground leave it there until the bees have gone up into the hive. All this should be performed early in the morning, and as gently as possible. Bees are a tender insect, very easily killed, and for that reason no hammering or rapping about the hive should ever be allowed. No deleterious smoke should ever be ap[p]lied. The smoke of burning cotton, blowed in for a little while, at the moment of removing the top, will be sufficient to drive the bees into the slack of your cloth at the other end, when the combs may be taken out without killing any of them. It does not damage a kin[g]dom of bees to take their honey bearing combs from them at the time the cornfields are in full bloom. They will fill it up again, in 8 to 12 days. I take their honey, from the 10th to the 20th of June, taking 2 or 3 every morning until I get through with my 16 hives: and that is as many as will do well at this point. Some places will support a much greater number. From these 16 hives, we get from 350 to 400 pounds of sealed up combs every June. We rob them but once a year.

It is not the small size of the moth fly that lets him into your hives. It is the honey loving avorice of the owner that opens the doors for him. He robs them too soon, too often, or too deeply; or, he may have too many hives, overstocking the honeyrange of his vicinity. . . .

Suspend your properly constructed hives on a rack, 2 or 3 feet from the ground, vent or outlet towards the south, and no less than 20 or 30 feet from any other hive; the more they are scattered the better. A bee house to crowd your hives in is not the thing for health and prosperity with the bee communities. Make your hives of good thick pine plank, shade them from the summer sun by laying a plank cover on top. Winter will not hurt them with such protecting hives.

Comply with these easily performed, simple rules at the right time, and the moth will have to seek their supplies of bee comb at some other place.

Lincecum then added:

P.S. Looking over this article, I find it too long perhaps. But it is a *sweet* topic, and somehow or other, my old, time jarred, mental machinery was not in a condition to make it shorter, or to put it up in better form. If you can make out to understand it, and deem it worth a place in the journal, you will oblige me by remodeling it to suit your taste and space. You may rest assured that with full swarms of healthy bees, that form of hive is full protection against every kind of insect; ants, moth, spiders, robber bees, and all other bee pests, except the avoricious, honey craving propensities of reckless man.

In his remarkable letter of June 19, 1861, to Dr. Dunn (which begins our introduction), Lincecum mentioned that his bees had rather free run of his office. In a second letter to Dr. Dunn (September 11, 1861), he elaborated on his intimate relationship with his bees:

You asked a question or two about the honey bee. The thermometer is now 94. For the sake of the cool south wind I sit in the gallery to write, being in the way where thousands of bees are passing to and from their basin of water on the floor a few feet from me. They occasionally, in their hurry, strike against me, sometimes becoming entangled in my hair or beard so badly that I have to stop writing to help them. And this circumstance brought to my mind your questions.

You wonder at the familiarity that subsists between the bees and me; that only shows that you have taken no points to cultivate an acquaintance with them. There are very few things in the animal kingdom, fool enough to be friendly with a man who would fight it with hat, fist, and coattail every time it passed near him. More especially the finely organized, sensitive little bee. He possesses a splendidly organized brain and is very capable of extensive training. I can set them to work almost anywhere I desire. They know when they find me far out on foot, in the prairie, which they prove by following and lighting upon my breast often. No bee ever lights on me when I am over two miles from home. Numbers of them sleep on the bed with me, and sometimes so many that I have to be careful when I move that I don't roll upon and kill the friendly little fellows. There are at this time sixty or seventy thousand of them in my gallery. I am constantly with them, and I teach them some useful lessons. They know nothing about swimming or helping themselves when they fall in the water. Thirty yards from my house is a large artificial pool. In this pool great numbers of my bees were daily drowned. To prevent which, I conceived the idea of trying to familiarize them with the water and thereby learn them how to avoid tumbling into it. Placing a basin on the gallery floor in which was a small quantity of water, when a number of them would be drinking in it, I would suddenly fill it up, and they being overwhelmed would rise to the top and make shift to scuffle out of the narrow basin. Repeating this experiment often every day, it was not long until many of these learned to be good swimmers. And now, when I fill up their water basin it disturbs them but very little. They will rise to the top, swim ashore and not fly away til he fills his water sack.

There are many more curious facts and powers about the bees. So there are about ants; who belong to the same order in Entomology [Hymenoptera]. The spider races and their natural enemies, the mud daubers, are full of interest, requiring careful and patient observation. To know all about every type of the insect world, down to the lowest microscopic animal scale, demands from the investigating mind, serious, earnest study. Man, having been progressively developed from an animal condition, cannot know himself until he is acquainted with the inferior types of the living and moving world: his kindred forms.

Among the many insects that found a home in Lincecum's office and the garden outside were a number of spiders.[12] In a letter to H. C. Wood, Jr. (June 22, 1866), Lincecum wrote:

We have scorpions very plenty—as great a variety of spiders as any country can produce. . . . From the red microscopic spider to the grand old hunter spider (*Mygale Hentzii*).[13] We have three or four of these large hunters; they all dwell in the ground in holes excavated by their own hands. Some of them do not spin! Others line their habitation to the very bottom with very soft white silk, and at the entrance he makes of the same material mixed with a little trash an ingeniously contrived trap door[14] hinged at one side, which he leaves open when he is out, but when he returns he shuts it down behind him, and seems to tie it within, for it cannot be raised without bursting something inside. I have seen them retreat into their hole, and the trap door seems to follow as the result of a vacuum, but the truth is, he hauled it down by a cord which he had prepared for the purpose, and which he had laid in a convenient place to snatch up quick, in case of a hurried retreat. There is another large spider here. When suspended in the center of his great circular snare or net he will measure from 1 to 3 inches from his front to his hind feet. The net is composed of the strongest kind of silk, capable of snaring our largest grasshoppers, and occasionally small birds are found tied up in their nets. The globular cases in which they deposit their eggs for winter quarters is a wonderful contrivance. It is near an inch in diameter, and the interior is interlaced in such manner with extremely fine silken threads as to suspend and hold the eggs in a stationary, but separate predickament. But I will enclose one of those egg cases, and you can examine it for yourself. These spiders are very numerous. They make their appearance in August; swinging up one or two of their great nets in every corner of one of the premises for miles, they continue through September.

On Christmas Eve, 1866, in a letter to Dr. E. T. Cresson, Lincecum wrote about his current entomological collection: "I have in my vials 21 spiders, 5 of which are aerial navigators—Balloonists." Anticipating incredulity on the part of Cresson, Lincecum hastily added: "They do positively construct balloons, take in passengers, and sail them, perhaps half across the continent in a day. I'll write you an account of these little balloonists some day." Lincecum did, indeed, write a description of his "balloonists." An article entitled "The Gossamer Spider" was published under his name in the *American Naturalist* in 1874.[15]

In a letter to Cresson's colleague, Elias Durand, December 24, 1865,

Lincecum wrote: "Almost every type of spider has his enemy type amongst the mud daubers. Even the great hairy tarantula (*Mygale Hentzii*) of Texas is often seen dragged along in a lifeless condition by a large red winged wasp of the dirt dauber type."

Next to the ants and bees, the dirt daubers[16] were Lincecum's favorite insects, and he gave them much attention. On October 7, 1863, he wrote to Dr. A. G. Lane[17]: "I have discovered 17 species of the mud dauber genus. . . . The mud dauber genus ranges from the great black-bodied Tarantula killer, till they are no larger than a sweet bee. Don't kill the mud daubers, for, was it not for them, the spiders would occupy every hole and corner of creation."

He was fascinated by the antagonism that prevailed between the spiders and the daubers. In response to a letter from his Austin friend, S. B. Buckley, Lincecum wrote (February 1,1866):

> Your description of our Tarentula and his antagonist is pretty good. You do not make the *Mygale* quite brave enough. He sometimes—not very often—succeeds in capturing the *Pompilus*,[18] and then as if to show off what a wonderful feat he has performed, he walks about sometime holding his great but vanquished foe in his mouth, carrying him at as great an elevation as he can streach his legs to perform. In 1864, I had a pet *Mygale Hentzii* in my garden. I fed him twice a day with grub worms; he would devour two large ones daily. I had also, in 1864, a very gentle *Pompilus* about the yard; he brought in a *Mygale* every evening about dusk for 4 months. He carried them under the dwelling house. At last he found my pet in the garden. I did not see the combat, but met him dragging the defeated tarantula out at the gate of my little garden. You say the same poison which *kills* the tarentula appears to prevent it from decaying. The spider is not killed, but paralyzed—it is in a state of suspended animation, and when kept in a dry place, is capable of resisting decomposition for many years. All that family of *wasps,* and I place the *Pompilus formosus*[19] at the head of them, have the power to paralyze their prey. I have found many times, the old time-worn daubbers nestes under shelving rocks in Georgia. They had been kept perfectly dry, but there would be found an occasional cell that had not been bored out, the egg that had been deposited in it, had proved abortive, and from the appearance of the wornout nest, had very probably been there 20 years. On opening the sealed up cells, the spiders were all perfect, limber and there were slight motion, seeming in their legs, and it may be possible that they were conscious of their deplorable condition. I should be afraid to let a large tarentula killer sting me. He might paralyze extensively. All that family of wasps are not spider killers. Some of them prey upon the cicada, one spe-

cies on the July fly, two species on the horseflies, and another on a large brown worm. . . .

I have never known of a case of tarantula bite, and cannot say whether he is as deadly as he has been represented or not. I think, however, they are not very dangerous. You say, speaking of the *Mygale*, "Providence has, to prevent too great an increase of these large spiders, created an insect of the wasp family," &c. That is to say that providence created the *Mygale* before he did the *Pompilus*! Well I never knew that before, but I give up, for on closer examination, I do think the spider, from his dusty appearance must be about one week the oldest. Then the question comes up, for what purpose did Providence create the spider? You never told me that you was acquainted with Providence before the heavenly war we have just passed through [the Civil War]. I was not in the war and did not get to see the mashed up bones and blood and ripped out hearts and guts, that took place so often, under the guidance as it is said of Providence or I might probably have made his acquaintance too. As it is, I must confess my entire ignorance of his characteristic traits, and shall not mention it in any of my notes on ants, scarabacus,[20] doodle bugs &c. until I am better acquainted with him.

On July 22, 1866, Lincecum wrote again to Buckley about the spiders and daubers:

The spiders except the hunters are not grown yet but I am watching and collecting statistics for their history. Their antagonists, the mud dauber family are in their prime now, and [I] have already collected quite a number of them. Among my tarentula killers I have the largest one I have yet seen. He measures from point to point of wings 3 1/2 inches, length of body 2 1/4 inches, from tail to point of the antenna 3 inches, from toe to toe of the longest legs 4 inches. . . . Doubtless, if it was not for the dirtdaubers, the whole earth with all its appertainances, would in a very short time be tied up and clothed in the folds of spider web.

There are also a very great number of this family or class of insects who do not prey upon the spider races, but [on] . . . cicada, catidids [katydids], grass crickets . . . no one of them is ever seen to offer, nor does any of the whole class ever capture any insect or spider for food for himself, but—and this is the characteristic trait that marks and places the whole of these insects in one class in my arrangement—for the purpose of hatching and nurturing the young. They all place their prey in some artificial or accidental cell, deposites an egg with it, and then very carefully seals it up.

In speaking of this family of insects, most of [the] people call them wasps, and lump them with the paper makers.[21] In their entire history the paper makers are very different. He constructs paper cells, deposits an egg in the naked cell, feeds it with honey after it hatches, which like the honey bee, he collects from the flowers, and as it grows older, he adds a little sausage meat, gradually increasing the quantity of animal food until the young one subsists on it alone. He makes his sausage meat, by cutting and mincing a nice smoothe skined caterpillar until he has a very soft pulpy ball of it. With this he feeds the baby wasp, and I dare suppose that while he is mincing the caterpillar, that he reduces the ball of sausage meat to the proper size for his aerial transportation by swallowing a few mouthfuls of it himself. The paper wasp dwell[s] in communities, and like the honey bees, have but one female in the community. Whereas the mud daubers observe the marriage institution, live and work together in pairs, and are never seen in communities. The paper wasp carefully nurses and feeds his young, while the others, after depositing the egg, pays no further attention to it. I shall never agree to call them all wasps.

Ever the curious observer and careful reporter, Lincecum described some other activities of daubers and wasps. In July, 1866 (undated), he wrote to Professor Wood:

Two or three days ago, I had picked up a leaf of a negus tree,[22] having on it a small jugshaped dirt daubers nest. I had often seen these little jugs sticking on green leaves before and sometimes on the ends of dried twigs, but had not seen, to know him, the builder of the jugs. For that purpose I had pinned up the leaf, and today, even now while writing the above, happening to cast my eyes in the direction of the little jug, observed something moving there, closer examination developed the fact, that the young dauber, was in the act of cutting his way out from the little jug, which he soon accomplished and was attempting to fly away when I captured him. Though I had seen him many times before, I had never accused him of being the jug maker. The jug is a globe, half inch in diameter, flattened a little on the side that adheres to the leaf, with a short neck on the top side. The insect that came out of it is one inch long, the thread that attaches the abdomen is very slim and rather over insect proportions in length, on the back of the head there is a yellow spot, another on the upper front of the thorax, two bright yellow bands and a yellow spot on top orniments the abdomen, his wings, are a very dark brown. When on the wing, very few people are able to distinguish him from the yellow jacket wasp,[23] of Texas. He is a spider killer and quite a bold one.

There is another, and as we do not understand the use of it, rather curious and certainly very cruel process, going on all around me, while

I am engaged in writing an account of it. It is being performed by the large red bodied, black winged, paper making wasp. It is a custom with them, and they do it every year, to commence dragging out their young in the full grown larva state, fly a little way with them, and dropping them, leave them to perish. They began this cruel work day before yesterday, and today there are many hundreds of the writhing white, fat larva strewed around on the floor. They come from the garret of my office, and as it is inaccessible, I am unable to say whether they are distroying their young to prevent a surplus, or whether they have war among the various nest[s] up there. I am inclined to the former opinion, as they do not drag out and distroy any in the pupa state. War would make no distinction.

Lincecum went on to say:

You wish me to give you as full an account of the symptoms following the bite of the centipede as I can, and ask me if I ever saw a case. Yes.

In the year 1850, I was called to a case six miles off, and when I arrived, the *little girl* was dead. It had survived the bite 5 or 6 hours. The whole surface was dappled over with livid spots, from the size of a dollar down to a five cents piece, and there was an elastic puffiness, giving the whole person an enlarged, or fat appearance. It had been dead but a few minutes when I arrived, and felt so soft and life like, that I examined it very careful, hoping I might find some indications that would encourage me to experiment a little; but the child was dead. —The case occurred in this wise. The mother, Mrs. Keene, was combing her hair; and dropping her comb, it fell through a crack in the floor near the wall. The little girl, who was about four years old, went immediately out to search for the comb, and running her hand through the fence rail underpinning, the Centipede,[24] who had been attracted by the comb, and had not had time to get away, saw her little white fingers approaching the place where the comb had fallen, turned, and made another plunge, for he is a beast of prey, and striking the childs thumb near the outer joint, run up towards the hand, leaving a track on the thumb, very similar to what a sharp, small spur would make were it rolled along on the skin. There were five little oosy holes made with the feet, and higher up on the thumb, the grab of the Caliper like mandibles was plainly to be seen. The symptoms were, according to the mothers account, instant complaint, which grew rapidly worse, which was discribed by the child as being all over her. Vomiting of pale yellow glary matter supervened, which continued at short intervals, with increasing violence, until the child, in a convulsive struggle ceased [to live]. . . . I could have made this account shorter, but not so satisfactory perhaps.

Five other cases of centipede bites, have occurred in this vicinity, none of whom died. The symptoms were the same as those discribed in the Keene child till the vomiting ensued; at this stage of its action, the pain and sufering was checked in four of the cases, in the fifth case it was checked before it had ran so far.

Earlier (June 22, 1866) Lincecum had observed to Wood:

Our centipedes are sometimes larger, and pretty plenty, distant from our houses. Near our dwellings the chickens eat them all up, and I think they are preyed upon by other animals, or they would be found more abundant in the woods at least. I have found them here occasionally 9 inches in length. They are all the reptile that I fear, because they can hurt you, by walking on you, when they do not intend it. When they grasp you with their strong caliper mandibles, it produces most excruciating pain, and without immediate, efficient help the patient soon dies. The root of the *Tephroid virg.*[25] is the only specific for their poison that I know of.

Several months later (November 8, 1866) Lincecum gave Wood a graphic description of the predatory habits of the centipede:

I have seen in the hiding place of a large *T. heros*,[26] what I supposed to be her eggs. They were eggs of some sort; of a greenish blue color, perfectly round, lying in 3 rows and about so large.

   oooo
   oooooo
   oooooo

This scolopendra[27] is a beast of prey subsisting principally on the black cricket; he will devour grasshoppers and any other insect he can catch. He delights in creeping into the garrets where he feasts on the larva and pupa of the dirt daubers, which he obtains by tearing their clay houses to pieces. I was standing near a little heap of broken rocks one day, and suddenly a great many crickets came hurrying out from among the rocks, and leaping wildly, hastened away in all directions. I did not wonder long about the cause of their precipitate flight, before his scraggy looking majesty, a very large *S. heros*,[28] made his appearance, rapidly running around and under the broken rocks, eagerly searching for crickets. He did not, that time, succeed in capturing any of them, for they, having got the alarm in time, did not cease to hop, nor stop a moment as long as they had strength to jump an inch, and when they could hop no more, they ran as fast as they could, till they were scattered in the weeds and grass, from their terrible foe, who did not pursue, for, ugly and poison as he is, he too has his enemies, and was afraid to leave the pile of rocks to pursue his game. The greatest enemy to the Centipede are the domestic fowls, Chickens particularly. When an old hen with a brood of chicken[s] finds a centipede, she warns her chickens to stand off, and pitches into the reptile.[29] The little chicks show the greatest anxiety to rush in and help while their mother is fighting and picking the centipede to pieces, but she by continual warnings and threatenings succeeds in keeping them out of danger until she has jointed every segment, when she swallows the head part herself, and calls her brood to the delicious feast, which they seem to enjoy as a rare luxury.

Being a physician as well as a naturalist, Lincecum took special interest in the bites of insects and arachnids. He advised Wood (October 8, 1866) that "But few spiders can bite to hurt much," and went on to say:

The most painful and most dangerous spider is not larger than a small grain of wheat; of a pale greenish gray color, and when he is frightened, moves briskly by limping along.[30] One of these small creatures bit Mr. D. Tidwell[31] early in the morning on his hand, it soon became seriously painful; he sent for me, and when I arrived at 10 A.M. he was in great distress, attended with incessant vomiting of a bile colored, glary slime. I treated him with most active antipoisons six hours before the pains gave way. It left him with his lower extremities paralyzed, from which he did not entirely recover in three months. I have witnessed two cases of this kind.

In the same letter Lincecum went on to tell Wood of his latest activity: "While we were out today, turning over an old log we discovered a pretty large sized *S. heros*, which we captured, not without risk, and battled him. He is 7. ins. I must fix some arrangement to send you a set of the small myriapoda from my collection, so that you may be studying them, at your leisure. I will send you a female scorpion this time, and if the vial will contain any more, 2 or three little leggy fellows." Before he posted this letter, Lincecum added a postscript: "9 Oct. [illeg.] P.M. Thermometer 78 degrees, very clear, and the canadian and white fronted geese, in large flocks, going south.—Well, I spent the greater part of the day rolling old logs and chunks and brush with but little success, having found but two scorpion skins, and they in a dilapidated condition. One of them is only the body part, the other which is entire, but very tender, is herewith enclosed."

One month later Lincecum sent Wood some live scorpions and

expanded on his belief that these organisms qualified as mammals.[32] He said to Wood (November 8, 1866):

> I do not know certainly, but from my knowledge of the scorpions at present, I am mostly inclined to the opinion that the species we have here, as I mentioned to you on a former occasion, are viviporous. They carry their young on their back.—So does many species of spiders.—But the spider also carries her sack of eggs about with her until they hatch, and then takes the young ones on her back, carrying a hundred or more.—The scorpion is never seen with an egg sack, or eggs any where else. I have turned up logs, boards, fence rails, rocks, thousands of times, under which I have observed scorpions, it seems to me, in all conditions, attitudes, and stages of existence that they could be placed in, and to this day have never seen anything I took to be a scorpions egg; but I have often found them with young ones clinging about their armpits and on their backs, and sometimes so small and white that it required pretty strong glasses to make satisfactory examination of them. They have four appendages on each side of the under part of their bodies, which are, at the time of carrying their young, full, plump, lubby like, and white on their distended tops. When they have no young ones, these appendages are flatted down, and barely visible to the naked eye. But I will investigate this limb of the subject more industriously next year.
>
> I will try to find something that will hold and protect 2 or 3 living scorpions to send with this letter. If it is cold weather when they reach you, you must not open them till you have had them several hours in a warm room. They are very tenacious of life and will recover when carefully thawed from an icicle. To keep them healthy however, for future observation, they should not be suddenly revived from a frozen condition. We frequently find them in the spring season, amongst the clothing that had been laid away all winter; they are active and in good condition, often stinging for a mere touch, when it is certain that they have spent the winter without food. The scorpion is carnivorous, feeding on many small insects. They are sometimes observed eating crickets. Under the hearths of our fireplaces is a great place for them to take up their winter quarters, & every warm spell during winter, they come out on the hearth, hunting something to eat.

To his delight and satisfaction, Lincecum learned from Wood that two of the female scorpions survived the long trip to the East Coast by mail in good condition. Lincecum had assured Wood that scorpions posed no real danger. In a July, 1866 (undated), letter to Wood he went so far as to say: "The scorpion is of no consequence at all. . . . He travels over all parts of our houses, frequently dropping down upon us from the ceiling, and stinging us in bed. It frightens the women and children, but in all my long life, it has never happened in the circle of my knowledge, that any serious injury resulted from the sting of the scorpion, and the scorpion I speak of now, is the true one." In another letter to Wood (June 22, 1866), however, Lincecum did acknowledge that a scorpion's sting could hurt. "They have stung me often," he wrote. "It produces sharp burning pain, not more severe than the bee sting, and lasting but [a] little while."

If Horatio Wood had but little to fear from Lincecum's scorpions, he had nothing at all to fear from the daddy-long-legs[33] he asked Lincecum to send him. In a letter dated November 8, 1866, Lincecum readily agreed to Wood's request:

> I shall collect all the daddies I find, but I mustn't look in old churches, or I might find a bushel at once. Same way in the large hollow trees, in the woods. The daddies are seen from early spring till Christmas. They seem to have about the same length of leg, but their bodies are not so large as the November daddies. I have never seen a young one, and know not whether they are oviporous [egg-laying] or otherwise. They sometimes settle for winter quarters in some old church or other out houses in incredible numbers. I have seen a knot of them hanging to one another, under the roof of an old log cabbin church as large as a peck measure. —The daddies of the woods are found at this season of the year, in the cracks of the bark of the post oak trees; seldom on any other species of tree. Towards evening, now the air is cooling down a little, they are seen clustering about the bunches of long moss (T. usneoidis). Thousands of them pass the winter in the swinging moss, coming out in early spring, looking as new and bright as any daddy long-legs at any season ever looked.

On December 6, 1866, he told Wood:

> Of the Daddy long legs, I have perhaps 100. I see no difference in the house species, and woods and rock kinds. There are the white bellied ones, which I think are the females, and those that have not white bellies. There are a shorter leged black species, and another of the same form that are red. These are perhaps, the males and females of that species. There is one thing which obtains universally with the true daddy long legs, when you catch one of them, it spirts out a small quantity of a milky fluid, having a strong, rather disagreeable, spicy odor. And the odor is exactly the same with all of them. The shorter leged species have no odor that I have discovered, though

one of my little entomologist[s] says they have a slight odor which resembles somewhat, the odor of the daddies.

Virtually every insect interested Lincecum, and he took a special delight in some members of the Coleoptera, the beetles. To E. T. Cresson in Philadelphia he wrote (July 16, 1866):

> . . . during some seasons, we have a considerable supply and extensive variety. — Coleoptera affords a wonderful variety, and some wonderful individuals. The largest specimens in this order are found in the water. There are many quite large water beetles, besides our great tyrant of the pool. I have one of these 2 1/2 inches in length, 1 1/2 in. in width over the back; 2 pairs of oars 1 1/2 in. in length.[34] Wings very fine, and when expanded 4 1/2 inches from point to point, sheathing coriaceous, wings straight, the edges of his oars beautifully set with stiff hairs presenting a fine effect while swimming in clear water. His color is a dark brown. Flies and swims both, very swiftly. There is nothing disagreeable in his external appearance, and all his movements in the water are surpassingly graceful. As he travels by night, I have never been so fortunate as to see him on the wing. They are rather a rare bug this far N. any how, and being very shy, they dart away and hide as quick as a fish when they discover any one approaching the water during daylight.
>
> Some of these water beetles are no larger than a big flea. While writing at night, I have observed these very small, green Boat flies,[35] having in their nocturnal flight, been attracted by my lamp, struck against it, and falling on the table. I have gathered them up, & pleased them very much by tumbling them headlong into a glass of water, which they manifested by their playful gambole and deep and frequent divings.

Some beetles could be more of the nature of pests than amusement, as he described to Wood (June 22, 1866):

> Our coleoptera is extensive. At my camp, of a night in May, I have heard the sound of the wings of a single species, so numerous that the whole weight of the atmosphere seemed to move and vibrate like a chord in music. We made up a large brushwood fire to give a light while we were preparing supper, but the beetles being fascinated and attracted by the great light, poured into the fire in such countless thousands, that they put out the light, and their roasting and frying carcasses produced such an intolerable stench that we were forced to seek another camp and go without our supper.

A more serious pest was a cabbage bug whose name Lincecum had not yet determined when he sent an article on the devastating insect to E. T. Cresson on July 16, 1866:

Herewith please find a short history of a Texas species of Inj[urious] Ins[ect] with a small bottle containing specimens of the perfect insect and some of their eggs.

Please after you have examined them and satisfied yourself in that respect send me the results with the order and specific name of the insect. —

They are a very injurious bug. Nothing is more certain than the entire destruction of every plant belonging to the order Cruciferae, that is found in your garden when they fall upon it, if let alone. I know not how far they range north. I never saw them east of the Mississippi river, and do not suppose they reach even that far east. People all around me say they have lost all their Cabbages entirely, two or three weeks ago.

It is a very easy matter while you are hunting your cabbage and other plants for the insects, to find their eggs also, and by destroying them, soon break them up for the season. I am certain that if my neighbors had all paid the same attention to their gardens that I have, that they would have saved their vegetables and been done with it long ago, for I have not seen a wingless bug in my garden in two months.

I have received the numbers up to June of your clever little monthly[36] and am sorry to hear you talk so disparagingly as you do in the June number.

Lincecum enclosed the following note to Cresson, and it appeared in the *Practical Entomologist* as "The Texan Cabbage-bug."[37]

Herewith, for your inspection, I send a set of dead specimens of a very destructive insect that I have seen nowhere, but in Texas. Mr. J. H. Hilliard says that he saw them in Louisiana, and that they were confined to, and destroyed his garden several times, in the same manner that they are sucking my plants to death. As here, he says the Louisiana bugs touched none but the cruciform plants.

Year before last, they got into my garden and utterly destroyed my cabbages, radishes, mustard, Seed turnips—every cruciform plant. Last year, I did not set any of that order of plants in my garden. But the present year, thinking they had probably left the primises, I planted my garden with raddishes, mustard and a variety of cabbages. By the first of April, the mustard and raddishes were large enough for use, and I discovered that the insect had also commenced on them. I commenced picking them off by hand and tramping them under foot. By that means I have preserved my 486 cabbages, but I have visited every one of them daily *now* 4 months, finding on them from 35 to 60 full grown insects every day, some coupled and some in the act of depositing their eggs. Although many have been hatched in my garden the present season, I have suffered none to come to maturity, and the daily supplies of grown insects that I have been blessed with, are imigrants from some other garden.

The specimens I send you will be better than an attempt at discription; I shall write only as to their history. The perfect Insect lives through the winter and is ready to deposite his eggs early as 15th March, or sooner, if he finds any cruciform plant large enough. They set their eggs on end in two rows, cemented together, mostly on the underside of the leaf, and generally 11–12 in number, — but I send you eggs specimens also. — In about six days, in April—4 days in July—they hatch out a brood of, except the wings, perfect little insects, colored spects and all, who immediately begin the work of destruction, by piercing and sucking the life sap from the leaf that fostered and protected them during the period of their incubation. In 12 days they have matured, and are found in pairs almost universally. I have often observed the half grown ones coupled. They are timid, and will run off and hide behind the first leafstem, or any part of the plant that will answer the purpose. The leaf that they puncture immediately wilts, like the effects of poison, and soon withers. Half a dozen grown insects will kill a cabbage in a day. They continue through the summer, and sufficient perfect insects survive the winter to ensure a full crop of them for the coming season.

This tribe of insects do not seem to be liable to the attacks of any of the cannible races—either in the egg state or at any other stage. Our birds pay no attention to them, neither will the domestic fowls touch them. I have as yet, found no way to get clear of them, but to pick them off by hand. Please send me its name and place in scientific classification and oblige.

Among Lincecum's many scientific correspondents and beneficiaries was George W. Peck of New York, who had a special interest in butterflies. He wrote to Lincecum in early 1867 asking for information on Texas butterflies and, hopefully, for some specimens as well. Lincecum replied most cordially on February 14, 1867:

Your very kind and very nice letter of 29th. ult. reached me yesterday evening. Low pitched developments never write such letters as it is. It is a great pleasure to me to receive such sensible familiar friendly papers. I may not be able to reciprocate fully, but I beg you to continue to write to me, notwithstanding. I am not at all surprised at the slightly manner in which you speak of the "poor closet naturalist." Yet there is a pleasure in breeding and raising the wild untaught infants to maturity, even in the closet. If however, you can take great delight in breeding them indoors, how much more delightful would it be—how much more vigorously would your caged up spirit essay to leap forth—could you strike out on the widespread prairies at the season when they are all bedecked with gorgeous flowers of every hue, that are waving their honey charged heads in the gentle south wind; while thousands and tens of thousands of your speciality, dressed in their gayest and brightest habiliments, are culling the sweets from the myriad tender heads of the blooming expanse? Oh! let me alone in the woods and wide untrodden plains. Let me remain with my "kindred forms, my brother emmets and my sister worms." But I cannot stop to discribe to you, the beautiful scenery of our bright, far reaching prairies, nor the soul inspiring influences they exert upon the intellectual be-holder. . . .

About the butterflies, though I am quite familiar with them I have written nothing, except an article for the *Texas Almanac*, for the present year, on the subject of the caterpillar.[38] As every one in the south is acquainted with the fly that lays the eggs, I did not discribe it, only gave directions as to the means of destroying it. I hatched and fed a number of them, and watched and noticed their habits through several changes from the egg to maturity. I sent the perfect flies with the large collection of butterflies to Prof. H. C. Wood jr. Acad. Nat. Science, Phila. I shall take my net . . . with me in my excursions and shall expect to find some new things in Lepidoptera, while in the fragmentary points of the rocky mountains that jut down into N. W. Texas. I shall trust the collections of the Lepidoptera of the plains, while I am in elevated regions, to my little class of Entomological schoolgirls. In this way I shall be able to make a large collection from [a] wide range of country, and towards the first of September next you may be looking for a nice box of preserved butterflies.

Before closing his letter, Lincecum added: "The inclosed moth, crawled up from the ground and sat on my door steps yesterday till its wings expanded, when I captured it. It contained 499 eggs, which I took out of it. I dried it partially by the fire."

Lincecum did, indeed, collect a number of butterflies for his friend, Peck, and wrote him as follows (November 19, 1867): "After capturing some Novr butterflies on the 15th, I prepared them, put them in the box and nailed them up. In that box is placed in white paper envelopes a thousand to twelve hundred specimens of butterflies, collected from 23 Texas Counties, and a few from Louisiana. There are specimens for every month, excep[t] January and July. December has not come yet. I hope the collection will be acceptible. I shall deposite the box at the Depot as soon as it is safe to go there. The yellow fever has retarded my progress very much this year. . . ."

Because he was ill with yellow fever, it was the ninth of December before Lincecum was able to take the box of butterflies to the express depot in Brenham. By that time the number of specimens in the box had increased to two thousand—or else, by that time, Lincecum had forgotten how many he had enclosed, for that is the number he cited in a letter of December 11, 1867, to Spencer Baird. His diary for 1867 has the following entry dated November 13: "I have nearly got my specimens ready for sending off. They are well preserved and in good order, and I am sending them to the Academies of Natural Sciences for the purpose of increasing the stock of knowledge among mankind. I am not sending them to men for their individual benefit, but to their charge and guardianship, for the coming generations."[39]

Considering what a prolific writer Lincecum was and the great delight he took in the beauty and wonders of nature, it is surprising that he did not write more on Texas butterflies. As can be seen in the following little gem to Horatio Wood (July 3, 1866), his relative silence on that subject is a pity:

What I am collecting now, are chiefly nocturnal insects, which are attracted to my lamp, while I am writing. They are mostly of the scale winged races, ranging from the size of a gnat, to a common candle fly or miller. I have no other way to preserve them, but in spirits which I fear will destroy their brilliant colors. I suppose that it is to escape the observation of "fierce libubilia"[40] and other types of that gluttonous class, that forces these extremely minute butterflies to seek their lovers during the night time. All those insects that are pleasant food for birds and insect hawks avoid travelling or singing during day light.

In the next paragraph of this letter to Wood, Lincecum described an insect that had him rather baffled:

We have an insect here which is called by almost everybody "Cowkiller."[41] I have heard it called by perhaps *two men* "Ant lion."[42] It is in shape like an ant, and as large as a drone honeybee, has no wings, and is never seen in company with his kind. He travels rapidly, seeming to be hurriedly in persuite of something, which he never overtakes. In the course of my long life I suppose I have followed one and another of these cow killers, put it altogether, 20 miles, and as yet have not seen them find anything, nor meet with another of his kind.

He is armed with a very long sting, which is very painful when he uses it. I have several times visited children who having been attracted by the beauty of his brilliant red velvet coat, with the black band, had caught them up in their hands, and had been stung by them. The pain was excruciating, lasting three or four hours, with considerable swelling. Always going off without any lasting injury. The cow killer being always full grown when I find him, and always from home if he has a home, I, of course, know nothing about his origin. They are not often met with. I have never found one with wings, yet from his form, wings seem to be indicated, and I think it quite probable that a degree or two further south, he may be found with wings. I shall place him in aptera[43] until he shows me his wings. Our cutting Ant, (*Myrmica Texana)* has a queen, or mother and, after she has cut her wings off, that is very much the size and shape of the cow killer, but does not adorn herself with the scarlet and black velvet as does the wandering cowkiller. I will say no more about this lonely insect at this sitting. I shall preserve a few of them. —

Lincecum then added: "I must have a work on entomology, or I can't get along very well with my bugs. Buckley wrote me the other day, that Harris' treatise on insects, On Entomology with plates, 2nd or last edition, Edited by Dr. Fitch, is the best work."[44]

Indeed, all of Lincecum's prodigious work on insects was hampered by his lack of adequate instruments, vials, preservation chemicals, scholarly books, and other "tools of the trade." And his greatest handicap was his isolation on the Texas frontier, so far removed from other scientists, except by the mail, which, be it noted, made this body of letters available to later generations.

Chapter 4

Creatures of the Air, Land, and Sea

I will do my best in the preparation of birds and small mammals;
I am, however, very fumble-fisted. . . . I shall do all
I can as it is; if I get greater facilities I shall do more. When the specialists
get through with the examination of the specimens I have already
sent to the Institution, please send me an account of them.

Gideon Lincecum

During his 1835 tour of Texas, Lincecum had some exciting encounters with birds, especially the shore birds, sandhill cranes, white pelicans, and swans, as well as prairie chickens, "turtle doves," and turkeys.[1] After he settled in Texas in 1848 and began keeping meteorological journals,[2] his entries included frequent references to the migrations of birds as observed from his Long Point post. For example, in March, 1860, he recorded that "the white-fronted Geese commenced passing N. on the 1st," and "the Swans, en masse, passed up N. at 1 A.M. on the 4th."[3] In the fall he noted their return: "the first wild geese passed South on the 8th" of October and the "First Swans passed S 6 1/2 A.M." on the 21st of November. In his voluminous correspondence he wrote surprisingly little about birds, but some nuggets appear.

In his February 17, 1861, letter to S. B. Buckley, Lincecum commented on that favorite Texas bird, the roadrunner: "Your trip to the Fredericsburg and Fort Mason countries, will be an interesting one. . . . you will find . . . the chaperell cock,[4] which to me is a great curiosity, particularly in his habits, and in his food; very often in summertime, when you find one of them, he will be seemingly, quite careless as to your pres-

ence, slowly walking with a lizzard in his mouth. You set your dog after him, and as for fast running, your dog soon finds that it is no go on his part."

In 1866 he began collecting bird skins and bird eggs for Spencer Baird at the Smithsonian Institution and exchanged many letters with Baird on the subject. The following comes from a December 13, 1866, letter to Baird:

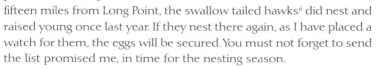

Herewith enclosed please find a receipt for three valuable books from the Smithsonian Institution. I am much pleased with them and will in some form or other reciprocate the great favor.

I have made arrangements with several hunters and herdsmen, and will see more on the subject, to collect the eggs of the rapacious birds next spring. I am informed by a hunter that in a grove of pin oaks (*Q. catesby*)[5] about fifteen miles from Long Point, the swallow tailed hawks[6] did nest and raised young once last year. If they nest there again, as I have placed a watch for them, the eggs will be secured. You must not forget to send the list promised me, in time for the nesting season.

Later Lincecum needed a drill and blowpipe to use in emptying the eggs for preservation of the shells, and Joseph Henry sent him these tools in March of 1867. They arrived while Lincecum was on his West Texas expedition[7] and he wrote Baird upon his return home as follows (July 8, 1867):

Your kind favor of 1st of March ult. overtook me at Austin as did also the instruments that accompanied it. I continued my excursion, and the weather there being unfavorable, did not attempt to reply until I had reached the falls of the Brazos, 10th April. I cannot now say what I wrote but I may say that if I wrote as I feel now, I returned my sincere thanks for the nice little instruments you were instrumental in procuring for, and forwarding to me. The lens is useful to me in many instances. The drill and blowpipe have been but little used. The spring was so uncommonly cold that few of the hawks nested at all. I did, however, from the mountains, find six swallow tailed hawks nests, and at the same time discovered the reason why collectors have not supplied the institution with the eggs of that gracefully moving bird. The nests of those I found were all in sight of each other, sometimes two on the same tree, situated on the pecans and cotton wood trees, and placed in the fork of the last boughs that are large enough to bear the weight of their small nests, and at least ten feet above the point where any human being capable of climbing such trees dare venture.

I very much regretted leaving them untouched, for I could plainly see that if I could obtain them, it would be a great rarity anywhere.

In a January 21, 1868, letter to Baird he describes his untutored method of skinning birds[8] for the Smithsonian and added a description of a rare sparrow:

I have made a trial at skinning birds, and find that I cannot succeed in a style that will be of any further use to you than to determine the species. You remarked in your letter that if I did not skin the birds, that you could determine the species from the head and wing if I would send them to you. My skins will be better than that, having the head, both wings, legs, feet and tail, and the entire skin though widely split. I never saw a bird skinned, nor any instructions on the subject. My method is to split the skin from back of the head, over the back to the root of the tail, draw out the neck bone, cut it off at the skull, ease out the wings to the last joint, the same with the legs to the knees, cut of[f] the tailbone with as little flesh and fat as possible, peel the skin from the breast, and the operation is completed. I then pour into the mouth and all the places where the bones have been jointed off a few drops of carbolic acid, stuff lightly the leg, cleaning the skins with dry cotton, spread out the skin, raw side up, as wide as I can, on a board and leave it there to dry. Having no means of extracting the brain, I fear it will, if left in the skull, spoil my work. You must send instructions for skinning birds and mammals and for preserving the skins; I have instructions on the subject of bird eggs, nests, &c. already sent me. —I do not think I can prepare skins that will do to *mount*. Send me the arsenical soap, carbolic acid, alcohol, &c. I have a drill and blow-pipe; no *proper syringe*. I will do my best in the preparation of birds and small mammals; I am, however, very *fumble-fisted*. Recollect that I am 75 years of age, and having been robbed, have nobody to help me.

I think it quite possible that I shall find some undiscribed birds here. There is a very small grass sparrow in our fields, that stands with the feathered race as the mouse does amongst the quadrupeds. Like the little field mouse, it is only found in the deep grass; running along its little pathes under the bending grass it is often taken for a mouse. I got one of them today, two of the Long spurs (Ploctrophanes) & one very pretty sparrow, with rusty red buts to his wings.[9]

Lincecum then added:

I went out shooting with Dr. Ruff, an industrious and considerably progressed young gentlemen, last friday and Saturday.[10] We skinned one burrowing owl, two prairie owls—these owls are 35 inches from tip to tip, and are only seen in grassy prairie, —One large light colored

Eagle shaped hawk—
He was 39 inches from tip to tip
legs long and muscular; mantilla
olive green to the nostrels, point
black; one butcher bird, one gnatbird,
one sapsucker, five sparrows,—2 species. One redbird
and one true snowbird.[11] The snowbird is rarely seen this
far south. I shall kill the migratory birds only during the time
between this and the nesting season. Then I shall take the birds
with the eggs and nests.

You must send me your catalogue of birds, and then I think I
shall have all the books I shall need about birds. I shall, if I get sufficient
alcohol, preserve the live freshwater and land shell[s].

I shall do all I can as it is; if I get greater facilities I shall do more.
When the specialists get through with the examination of the specimens I have already sent to the Institution, please send me an account of them.

During his January, 1868, expedition to the Gulf, Lincecum encountered the now endangered whooping crane,[12] but, as he says in a letter
to his son-in-law George Durham (February 16, 1868), the cranes were
in little danger from him: "Except the aquatic, I did not see many rare
birds during my tour on the Coast. I saw several pairs of the whooping
cranes, and thinking of your desire for a skin or two, I made an effort

to shoot some. Well, I didn't kill any, but speaking in the bounds of reason, I think I came in two miles of it." Also, in his February 16 letter to Durham, Lincecum said, "I have already sent off to the Smithsonian this year, a box of birdskins, and another of coast shells and some fossils. I believe I told you that I had sent all my ants and the history of what I had seen of their manners and customs to Prof. Cresson of the Entomological Society, Philadelphia," and he added: "The snipes have all left the Flag Pond. San's Duck was a bone fide, broad bill'd, flat footed, short legged duck but the squealing discribed by San might go by another name had you have hear[d] it.[13] If you recollect San wears an unpracticed musical ear."

On April 1, 1868, Lincecum reported to Spencer Baird on his continuing efforts to collect birds for the Smithsonian. He said:

> I secured another grass sparrow today, but I do not think it is like the choice one I sent you with some other birds not long ago. It is nevertheless a true grass sparrow, running mouse like and hiding under the bending grass. I trailed it, watching its creeping actions 100 yards or more before I shot it.
>
> Besides the grass sparrow, I collected one curlue and 2 plover skins.[14] The wrens only have began to lay their eggs. I shall be able to procure several species. It will not be long now till I send you some skins.

On the first of May, 1868, Lincecum wrote again to Baird, acknowledging the receipt of three pamphlets Baird had sent him. In a postscript dated May 5, 1868, he added:

> I have this day packed up what spec[imens] I had on hand, consisting of bird skins, Butterflies, some heads, birds nests, a few eggs, and a swamp rabit skin. Nothing new, but in the perpetual business of making preparation for my Tuxpan Journey,[15] I had not time to collect more. Indeed, there are, on account of the cold, wet spring, as I suppose, very few birds have made nests this season. The young men who will remain here after I am gone were finding some bird nests today, some had one, others two, three, &c. I hope they may succeed in making large collections. I shall turn over for their use the paper, Arsenic, &c that you sent me. Also my nets and other little fixings. I cannot say now what I may need in Tuxpan. I shall find out after reaching the beautiful valley. There are perhap[s] no birds in that country that has not been supplied to you long ago. It is among the ants and other insects that I expect to find rare things.

Because Lincecum was leaving soon for his new home in Tuxpan, Mexico, he sent Baird's work on North American birds[16] to his son-in-law Durham, who was living in Austin. On May 23, 1868, he wrote Durham that he had mailed the book and added:

> I am sure you would be much pleased with Prof. S. F. Baird of the Smithsonian Institution. He is the As[s]t. Sec. and one of the most

ardent devotees to science, particularly your specialty, and besides he is the most attentive correspondent among them all and a thorough gentleman.

Send him a few bird skins. Right at this time *our* sparrows interests him very much. But any bird skin with their eggs and nests are very welcome and desireable.

Unfortunately, Durham, who was born in England, insinuated in his response that American ornithologists like Baird were not in a class with their British-trained counterparts. Lincecum was no Anglophile, and he fired back with a hot, sarcastic letter (dated April 30, 1868, but placement in letterpress suggests May 30):

In your reply to my letter notifying you of the fact that I had laid aside for your use Mr. Bairds work on American birds, you gave me a sufficient hint if I had noticed it, to have deterred me from any further action on the occasion; nor should I have recommended Mr. Baird, the American Ornithologist, as a fitting person to correspond with you on such subjects. I regret very much that I have fallen into such an egregious blunder. It is not common for *me*, but you must recollect that I am getting a little old and with the hope that you will pardon this my first offence of the kind, I promise that I will not again insult the dignity of your European prejudices by the obtrusion of any more such *American* trash.

If it is very much in your way, you can, by some favorable oppertunity—some traveler—send it to my grandson George W. Lincecum,[17] who has just commenced the study of *American* birds. It will doubtless please him, and will, when he understands it, be sufficient distinction for a native born Texan among his kind.

Lincecum did not take as much interest in quadrupeds as he did in insects and birds, though as a youth he had been an avid and expert hunter of mammals, large and small, partly for sport but mostly for food.[18] After the Civil War he sought to furnish his eastern scientific friends with specimens of Texas mammals, either skins or bones or even, on rare occasions, some that were very much alive. He urged some of his Texas friends to collect skulls of various mammals, including those of American Indians, though by no means did he suggest that the latter be killed for this scientific purpose. In his letter to Solomon Cox[19] of Belton, Texas, January 12, 1867, Lincecum stated:

Mr. S. H. Summers gave me your name, and he recommended you as a man who, if any one in the state could find an Indians skull, it was yourself. His statements to me, of your peculiar travelling habits, encouraged me to write to you and ask you to aid, as far as suits your convenience, in making collections of Indian skulls. Any number, from one to one hundred, of whole skulls will be very acceptable. If you get the skulls of different tribes, please number them, set the name of the tribe with the numbers on a piece of paper, which, with your name, will accompany the skulls to the [Smithsonian] Institution, where together, they will make up a part of the great scientific work on the skulls of the American Indians.

During his three-month exploratory expedition to West Texas that spring, he gathered many skins and bones of mammals and heard at least one cougar. To his friend Durand (June 27, 1867), he wrote:

I had a pleasant time during the 3 months I was camping out. Having two young men with me, good hunters, who kept our larder well stored with various kinds of wild game, fish and the like, we lived a little better than home folks. I often wished you had been with us. I think the loud, wild and totally indiscribable, midnight scream of the great mountain couger would have delighted you. They are a sort of vantriloquist. They can throw their voice in such manner that you would suppose them to be on the top of the nearest hill, when in reality, they are three or four miles distant. This delusion frightens and sets in commotion all smaller beasts, who, not knowing where the sound comes from, are as apt to run to him as any other direction, and thus he takes his easy prey.

The following year he shipped to S. F. Baird some live, tiny mammals for which he had no name, but his letter (February 24, 1868) included a vivid description of their ferocity.

I ship today, to your care, a small box, containing a pair of full grown, living mammals,[20]—quadrupeds—I have put in the box crickets and bread enough to last them through, I think, if there should not be long delays. They are insectivorous, and from the signs in and about their nests, they feed principally on crickets and grasshoppers. Since I have been feeding them, they have learned to eat a little cornbread and to nibble a biscuit. They drink water freely and often. They will eat the lean flesh of a rabbit, or bird. I hope they may reach you in good condition. . . .

I had a family of them and fed them a week, where I could observe all their actions. I had the father and mother and their four half grown offsprings. They were pretty pets, and I had hoped to succeed in sending the whole family to you; but our highest, cherished hopes are often frustrated. The male made his escape, and finding another newly married pair—they too marry and as far as I can learn stick together as long as they both live—I put them into the box with my half civilized family of them. The male instantly caught a young one, and was aiming to kill it, when I put him and his companion into an empty oyster can, and setting it back in the box, went to supper. When I returned I found that the ferocious, rascally male had made shift to get out of the can, and had murdered all the young ones. I was very sorry for the loss, and thinking he had done all the mischief he could, turn[ed] his wife out of the oyster can and left them in the box with the bereaved and deeply afflicted mother. Next morning, I found they had murdered the sorrowing mother, and had eat her very nearly up. It is the two last captured cannibals, that I have sent you.

Unfortunately, this ferocious pair did not survive the trip, but upon hearing the news Lincecum retorted to Prof. S. F. Baird, April 1, 1868: "I am very sorry the birds box of mice was a failure. I'll try again. They are rare in this country."

On more than one occasion, a letter that Lincecum wrote to one of his scientist friends detailing some of his observations was subsequently published in a journal, and he was credited as the author.[21] Following are his descriptions of the opossum and the gregarious rat that appeared in the *American Naturalist*:[22]

The Opossum. —This species of marsupial, seems to be widely distributed in every portion of the United States. Its original name in the Choctaw language is "shookhutta"; which signifies that he is the father or rather the originator of all hogs. It is not very swift of foot, neither is it very wild. I have frequently, when hunting in the woods, passed within a few steps of them and they did not seem to regard me. Our turkey buzzards have somehow found it out, and will alight near where they find the opossum feeding in the woods and running up on him, flap their wings violently over him a few times, when the opossum goes into a spasm, and the buzzards very deliberately proceed to pick out its exposed eyes and generally take a pretty good bite from its neck and shoulders; the opossum lying on its side all the time and grunting. I have twice seen a buzzard do as described, and once I found a poor creature trying to find something to eat with one eye out and one shoulder entirely gone, evidently caused by a buzzard.

They dwell in hollow logs, stumps and in holes at the root of the trees. They do not burrow or prepare dens for themselves, but find such as are ready made. I have seen them carrying into their holes, at the approach of cold weather, considerable bundles of dry leaves rolled up in their tail; they understand the signs of the coming spells of bad weather, and they prepare for it by making for themselves a good warm bed. They do not hibernate, but are found out hunting food in frosty weather. They possess but little caution. Hence they are often found in the poultry houses, chicken coops, smoke houses, and even in our dining rooms rattling about in search of something to eat. I have often seen their tracks in the roads and paths where they had travelled three, or four miles to a farmyard, to which they had no doubt been directed by the crowing of the roosters. They will catch a grown hen and drag her off squalling at the top of her voice and will not abandon her until the dogs which have been aroused by the up-roar have overtaken and commenced cracking their bones. They will eat bacon, dry beef, carrion, any kind of fowl, rabbits, any sort of small game, almost all the insects, and fruits of every variety. They vora-ciously devour the muskmelon, and several species of mushrooms; in short they are nearly omnivorous.

The only case in which it manifests any respectable degree of cautiousness is when it is hunted at night in the forest; on hearing the din and noise of the hunters, it with some difficulty makes shift to climb a small tree or sapling, where, wrapping the naked rasplike tail around some convenient limb, it quietly awaits the approaching dogs and hunters. By many people the flesh is considered delicious. In Galveston, Texas, in the proper season a good fat opossum will sell for $1.50. Its flavor resembles that of the flesh of a young hog, but is sweeter, less gross and is no doubt a more healthy food for man. A dog will starve sooner than eat the flesh of an opossum; negroes and many other persons are exceedingly fond of it.

During their rutting season, the males are very rampant and bel-ligerent. Numbers will collect around a female and fight like dogs. Twenty or thirty years ago, I witnessed a case myself in the forests of Mississippi. The female was present, there were three males, two of them were fighting, while the third was sitting off a little piece, look-ing as though he felt as if he had seen enough. They were fighting hard and had been, from the signs in the wallowed down grass, for three or four days. Kicking over the female, who immediately went into a spasm, I made a slight examination of the pouch.

They are exceedingly tenacious of life. I have many times seen dogs catch them and chew and crack, seemingly, all the bones in the skin, leaving them to all appearances entirely lifeless; and, going out the next morning for the purpose of removing the dead thing, would find that it had left its death bed and putting the dogs on its track trail him a mile or more before overtaking him. He would, to be sure, be found in a bad fix; but at the same time he lacked two or three more bone crackings of being dead. They cannot, like the raccoon, be so far domesticated as to form any attachment for persons or their houses, though I have two or three times found them under the floor of dwell-ing houses, where they had been for some time and had evidently taken up winter quarters, but they did not remain there long, nor do I think they dwell long at any one place. They swim very well when it is necessary.—Gideon Lincecum, *Long Point, Texas.*—*Communicated by the Smithsonian Institution.*

Another of his letters that appeared as an article in the *American Naturalist* illustrates his tendency to anthropomorphize in a manner that made academically trained scientists distrust his observations.

The Gregarious Rat of Texas *(Sigmodon Berlandierii)*—This is a burrowing, gregarious rat, and like the Prairie dog lives in towns on the prairie. They dwell together in families. They prefer light sandy soil on the prairie, where the shivered limy sandstone crops out, but when the prairie is enclosed and cultivated, they take possession of the fencing, and burrowing under the bottom rail, excavate sufficient cells and construct their copious grassy beds there. Out on the prairie, in the wild state, they make one principal burrow, in front of which they pile up the earth that comes from all their subterranean galleries. They rarely extend their main burrow more than eight or nine inches in depth while their underground passages are seldom more than four or five inches below the surface. They also construct several secret outlets, opening ten or twelve inches from the main hole, which openings they very ingeniously conceal by strewing a few grass blades over it; and so, when the rat hunter attacks the citadel, the inmates escape through some of the concealed passages. Eight or nine inches deep and turned a little to one side in the main hole is a cavity seven or eight inches in diameter, filled with fine, soft grass blades, which must be quite warm and pleasant, serving the family for winter quarters. During the hot months, they construct nice grass beds in a basinlike cavity, which they dig out, under the sides of large tufts of grass, or little heaps of brush. The above is about the average customs of the distinct families in reference to the manner of making their homes, and in the same district in suitable soil, they construct many such family residences, and cut out very nice, clean roads from one to another in all directions. The grass, weeds, dewberry briers and everything in the way are cut out and carried away leaving the road about two inches wide, underrunning the grass and other rank growths that may fall in the way. I have traced some of these roads fifty or sixty yards upon which there had been so much labor expended that it could not have been the result of individual enterprise. These roads, which bear the indications of much travel, are evidently the result of a unanimous governmental effort. They are found universally in their cities, and passing from house to house there are many cross roads.

This Rat has a large thick head, nothing remarkable about the mouth and nose, eyes full, black and lustrous, ears half of an inch high and nearly circular; neck very short, body short and large; tail three and three-fourths inches long, clothed with very short, thick set hair; feet with five toes, nails strong. No cheek pouches; no grooves about the incisors, not very long hairs or "smellers" on the nose. Coloration a brownish gray. —G. Lincecum, *Long Point, Texas.* — *Communicated by the Smithsonian Institution.*

Lincecum sought to provide his scientific friends specimens of Texas reptiles but had to disappoint them for the most part when it came to snakes. To Prof. Horatio C. Wood, Jr., of the Philadelphia Academy he wrote (undated, July 23–29, 1866):

I notice Prof. [M. D.] Cope's slip of paper and am sorry that I shall not be able to do anything for him. This is not a snaky country. Formerly we had a great variety of snakes here. They are gone, and I attribute their absence to the long sore drought we have just passed through. By the by, what species of animal does Prof. Cope call salamander? In Ga. where I was raised, they called a species of the genus *Talpa* Salamander. In Texas they call the same mole gofer [gopher].[23] If by that term Prof. Cope alludes to what in Texas is called the "horned frog,"[24] but in reality is a short tailed, spiny lizard, having no trait of batrachian character about him, I shall understand him to include in the term *salamander* all our sand lizards, and we have a great many species of them.

In November Lincecum received a friendly letter from Prof. Cope (who, Lincecum said, revealed himself by his style to be a Quaker).[25] Lincecum readily consented to Cope's request for some specimens from Texas lakes and streams, including salamanders as well as fish. Lincecum's reply (November 18, 1866) to his Quaker correspondent uncharacteristically used "thees" and "thous."

My Dear Friend:

Thy easy, familiar, and quite friendly communication of 5th inst. reached me in eleven days. I am pleased with its contents, and will as soon as our streams get water in them again, and the fish comes, which will be towards the first of march, endeavor to make some collections, which I will send thee.

We have little lakes, containing perhaps 150 acres, they are deep and do not dry up. In these lakes there are some of the largest fresh water fish, that I ever saw in any country. The largest I have seen had a head and mouth very similar to that of an alligator, whole length of the fish about 8 feet, and would weigh probably 150 or 200 pounds. It belongs to the gar family, and is quite plenty in our lakes. They are

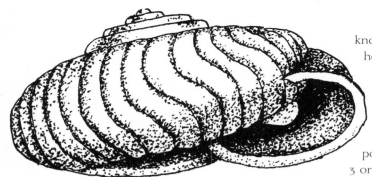

known to the people here as alligator gar. We have the buffalo fish weighing 25 pounds, blue and yellow catfish 15–18 pounds, black bass 3 or 4 pounds, with a considerable variety of small fish, belonging almost exclusively to the perch family.

The lakes, streams, and coastal beaches provided Lincecum other specimens that were of interest to his scientific friends, and he furnished them a number of shells, including a fine collection gathered by his wife, Sarah,[26] during their camping trips to the Gulf coast. In a letter to John H. Douglas of Mason City, Illinois,[27] dated January 14, 1860, Lincecum's pen waxed eloquent as he described the rich conchology of the Gulf beaches and his wife's delight in gathering its treasures:

> Having our encampment immediately on the sea beach, where the salt spray dashes into your tents, and the eternal roar of the great deep frightens the uninitiated. Yet [behold] the clean-washed sandy shore with its million varieties [of] glittering and gorgeously painted shells of every shape and form that imagination could picture. Here, on these crackling and thickly scattered konks and clams, sun cockles and twisted helicis, with the foamy waves dashing, surge after surge, at their very feet, the curious lady [Sarah], enchanted with the rich prospect which lies so profusely strewed far and wide around her, strolls all day gathering and selecting the finest specimens, until her overloaded apron lap get[s] too heavy for further progress—when she is forced to sit down on the clean, dry sand, and cull and reselect from her collection the most sightly; reluctantly she abandons the refused surplus, and again begins the tedious search. All day she traverses the bespangled shore, until unwelcome twilight overtakes her far down the sea beaten beach. But not alarmed she hurries home, and there displaying her well earned trophies of the vasty deep declares herself that she is not right sure that she has not left a great many lovely shells behind, promising herself that tomorrow, if the day should not again turn out to be too short, she will be able to make a more beautiful selection.

A year after Sarah's death, Gideon donated her collection to the Smithsonian, as he reported to his friend Baird (in a February, 1868, letter that is undated): "The shells were picked up by my dear old departed one, on the Gulf shore, immediately south of the mouth of the Bernard. The collection was at first large and consisted of a very full asssortment of the shells of that particular point; but the chaffy-minded, light fingered gentlemen, and ladies too, have handled them too often. All the large and handsome specimens have been carried off. That, however, is probably better than to have lost the little unsightly shells, as few traveling collectors pay any attention to such."

Fossilized remains of shellfish and other marine life were abundant in most parts of the Texas terrain, which had once been a great seabed. Fossils had long interested Lincecum,[28] as had indeed the whole broad field of geology, and they helped to feed his interest in evolution and progress, as reflected in this letter of May 15, 1858, to Dr. W. Spillman, of Columbus, Mississippi:[29]

> I have in my possession a beautiful collection of fossil shells, and casts of many species of shells, which I consider very rare, numbers of them perhaps are new and undiscribed species; all of which were collected while traveling in the mountainous portions of middle Texas. There is in the bluff of a creek, about 30 miles above Austin, west of the Colorado, a stratum of shell casts, 3 1/2 feet thick and cropping out half a mile in length; underlying a stratum of soft, bluish, slaty clay on top of these is superimposed a stratefied, mountain elevation of at least 300 feet. On top of all this is a pretty considerable drift of water worn, flint pebbles of various sizes, from a pound or two, down to coarse sand, which are composed of perrywinkles, clams, and a great variety of species of what I take to be fresh water productions. . . . The bluff I have here attempted to describe, besides its wonderful stratum of shell casts, is made up of various strata of shells in a perfect state of preservation. I have oyster shells and some of the smaller griphia, which look as fresh & as bright as if they had only been dead two or three years and some of them with the valves still hinged together. . . .
>
> The greatest wonders are to be seen in the primitive sea beds, that form the base of some of our mountains and the overlaying strata—the subsequent formations, which still maintain their original position, and by their well defined and numerous strata declare themselves to be the progressive work of countless epochs of vast intervals of time.

Chapter 5

Texas Geology

I could consume a lifetime in the investigation of the wonderful,
and endlessly varied secret creations that are rolled
to the fretted beach by the ceaseless throbbings of the heart of
old OCEAN. Why really, it seems to me, to be the matrix of creation,
and that all our dry land animals and vegetables have crept
from beneath its waves.

Gideon Lincecum

Lincecum's written accounts of his first visit to Texas in 1835 indicate that he paid attention to geological formations and mineral deposits at that time, and the letters he wrote after he settled in Texas contain a number of references to his continued interest in geology. One of his earliest surviving letters dealing with science was written in 1855. It gives a detailed description of several aspects of the geology of Texas, indicating that Lincecum had done considerable exploration since moving his family there in 1848. The letter, dated September 23, 1855, is addressed to Dr. W. Spillman in Columbus, Mississippi:

I am seated by my writing desk, in far off Texas, engaged in the more rational and agreeable business of discribing to you, as well as I can, some of the interesting facts which I discovered during my late excursion in the nearest group—to my place—of volcanic hills and little mountains,[1] lying a few degrees N. of W. on the Colorado river and distant 150 miles. This river rises in the vast plains above, falls upon the above noticed group of mountains, and having had *sufficient time*, with its solvent and whetting powers, has succeeded in cutting a passage through them. This passage is not as yet quite finished, which fact is established by the still existing cascades and small cataracts where the river makes its way through the marble range. The floor of the river at these falls is worn into many channels through which the water passes, and these channels are so regular that a footman may, at low water, cross dry footed by stepping from ridge to ridge that rises up between the channels. Showing that ultimately this solid marble flooring will be worn away, when the river will roll its clear blue waters sea-wards more quietly.

The place I have reference to above is called the Marble falls;[2] a company of gentlemen have recently surveyed off a town at that place, and call it the Marble City. Here the water power is perhaps equal to any on the continent, and already the above named company of gentlemen have organized themselves into an association for the purpose of erecting cotton and woollen mills on an extensive scale. It is situated about 50 miles above Austin, is very healthy, producing wheat in its greatest perfection, and as a country for the growing of wool, it is not surpassed on any portion of the globe. But I did not sit down to write you an account of the production and extraordinary fertility of our large and most beautiful State. It was to give you an account of some of the cretacious fossils, fossilifirous formations and the mighty changes which—by fire and flood and countless epochs of time— have been produced in the geology of the portion of our State over which I traveled during six weeks of the past summer. I had no tests or instruments with me, and on that account shall only be able to give you a partial discription of the rough external of the country alluded to.

That portion of the State in which I reside—(Lat. 30° 16′ N., 100 miles from the gulf, and equidistant between the Brazos and Colorado rivers)—is a very productive, black prairie soil, underlaid everywhere at the depth of 25 to 30 feet with a coarse, whitish, stratified calcarious sandstone. This with the black soil above, make an average of the limy formation from 20 to 100 or 200 feet, with no marine fossils discernible to the naked eye. Underlying this is the green sand stone (very closely resembling that which shows out, in the banks of the Tombecbee about Columbus, Mi.).[3] On the top of this sand stone, which has evidently once been a dry land surface, and at that time lying within the tropics, is to be found many species of endogenous plants (now flint). Some of them quite large, profusely scattered with the exogenae, sometimes lying cross[wise] and [in] piles and in great numbers. Every place, where the deep water cuts on the slope from the prairie towards the Yegua[4] have deepened their channels down to the sand stone, the above named flinty fossils, are to be found in abundance. No animal remains, except those belonging to the big bone period, and they seem to have slided into the gullies from the surrounding elevations.

Leaving my place we will travel North west and west 300 miles. Very near (2 miles) we come in contact with a very considerable pile of igneous rocks, they are heaped up in a vall[e]y of limy soil immediately at the edge or N. W. termination of the first calcarious formation beginning to count from the gulf, lying in a direction from N.E. to S.W. for 12 or 15 miles in length and are elevated in some places to a highth of 150–200 feet. The rocks have been partially fused and show clearly that it has been the breaking up of many strata of various compositions. Among these rocks are to be found in considerable quantities rocks closely resembling the french burls [beryls], and the people have converted a good deal of it to the same purpose. There are also quantities of petrified wood, some of which having been fused is now glass, still retaining the concentric rings, knots, and sap pores. The violence which was exerted in this upheaval was sufficient to break and tumble the rocks together so that the order of the different strata are not at this [time] cognizable. It contains no perceptible animal remains. North westerly from these granite knobs, sets in the country that was originally called the cross timbers. This is a stripe of timber mostly post oak and cedar, lying nearly N.E. and S.W. and is in width from 30 to 40 miles. This has no lime, but abounds in Silex, Silecious clay, magnesia, iron, alum, some coal, various other minerals, and whole forrests in petrification. These last lie on and near the surface. Above this timber stripe, sets in another calcarious stripe, but this is finer and harder than the limestone below the timber. Some of it [is] marble of a very fine quality. Some specimens of which I shall send to you in my next remittance. Through all the marble region the surface of the country is very rough and broken. They are called mountains; well, they are little mountains, having the appearance of potatoe hills in the distance, and rising above the level of the river from 500 to 1000 feet. It seems to my mind, that the whole country was pretty level when it was first lifted up, and has since been, by the action of time and the water, wheted away into the condition we now find it. These little mountains are still daily diminishing in size. Every rain carries something down hill and there is no action to carry anything up again. From the base to the top of one of these mountains I counted 26 distinct rock strata of from one to three feet in thickness. Between each of the rock strata is a clayey deposit of from 30 to 50 feet, containing marine recrement; in some of them great quantities, specimens of which you will find in the next box I send you. Up the sides of these mountains, halfway up perhaps, crops out a stratum of blue limestone, very hard and rough on the surface; the roughness, as you will see by the specimens I send, occasioned by the increnitic deposits of which they seem to be mostly composed. Immediately on top of the dark colored lime stratum is sparsly scattered, pretty large fragments of slightly water-worn whitish rocks, which seem to be entirely composed of encrenites. It is an exceeding strong rock, requiring great violence to break it; I send you some pieces of that. On top of this again are various strata of other composition, principally cretaceous interstriped with sand and clay; and on top of all, flinty pebbles and small boulders, containing shells of various species. You will find them in the box also. Progressing up the Colorado we find the elevation greater, the volcanic violence has been much stronger, raising the marble beds in scattered fragments to a greater highth. Beneath which, we find a whitish, limy clay containing small shells, and this is based on the slate. Underlying all this is the granitic rocks, all cracked and shivered and displaced, showing that it had been torn with great violence, by force from below. All this is exhibited in the cliffs of the deep water cuts made by the tributaries in their course to the Colorado. The bottom of these streams, as well as the Colorado itself, are all thickly strewn with flint boulders of various sizes, from 40 pounds weight down to the finest sand. In all this district I saw no coal.

In the bluf of Barton's creek,[5] and 100 to 200 feet below, the surface of the surrounding country is a stratum 3 feet underlying a bed of slaty clay, composed entirely of the casts of many varieties of shells. So closely cemented together that it was difficult to get them out whole. Indeed, I could no[t] procure a full set of whole specimens. This stratum lay in a perfect horizontal line, and was visible for half a mile. It was underlaid with 10 or 12 feet of white limy clay, and that again by a sheet of conglomarate which formed the bed of the creek. These aquatic fossil casts are so numerous, and some of them so large and so closely packed together, that when we consider the fact that the direction of the drift by which they were thrown there is directly oposit to the course of the running water now and the superincumbent mountain, stratified as it is, with many stripes of animal and vegetable formations (marking as they certainly do, vast periods of illimitable duration), we stand mute in the confusion. And the cosmogony of the ancients, aye, and moderns too, dies out from the mind like a shooting meteor.

Convinced from his own observations that Texas rested upon a wealth of mineral resources, Lincecum was a strong supporter of the Geological Survey that the Texas State Legislature authorized in 1858 and appropriated money for.[6] When an abortive effort was made three years later to cancel the survey, Lincecum wrote the editor of the *Galveston News* (February 14, 1861) as follows:

> From the newspapers I discover that during the late extra session of our Legislature, an attempt was made to abolish the State Geological Survey. We do hope that the frenzy of party spite may not further meddle with this matter, and that the survey may be permitted to go on uninterrupted; because now more than ever, the resources of Texas need to be developed. Let the future be as it may, we wish to know and understand the agricultural resources and mineral wealth of the state. The survey will, if no more, give direction and stimulus to individual enterprise; teaching us where we may expect to find useful ores, coal, and other valuable minerals, as well as where they

Gideon attempted unsuccessfully to drill an artesian well at this site on his farm.
His letters indicate he had reached a depth of 200 feet when early fall rains broke the drought.
Photo by Keith Carter

are not to be found. The late discovery by the State Geologist (Dr. Moore) of an Iron mountain[7] in Llano County, equal in quality and quantity, as I understand, to the celebrated Iron mountain of Missouri, is of more value by far than the entire cost of the survey. It is known that the Iron mountain of Missouri gives millions of wealth, not only to its own state, but to all the states of the upper Mississippi. What would the great state of Pennsylvania be without its treasures of Iron and Coal? We say let the Geological survey be continued by all means.

In 1860, Long Point and the surrounding area was in the midst of a multiyear drought. Lincecum began the task of drilling an artesian well to provide water for his family, crops, and animals. The following letter to Caleb Forshey (April 6, 1860) pieces together Lincecum's thoughts on the drilling of the artesian well and the science of geology.

In answer to your questions, in regard to the Artesian I am working at: I expect to procure a gushing sluice of pure water. The Geological testimony upon which I base my enterprise, are the facts, in the first place, that our geological formations are very similar to that about Columbus, Mi[ss]issippi]—fossils, stratification and all; and there, they obtained artesian everywhere. In the next place, the country being dry on top, it argues that there is plenty of water below. I am a believer in the doctrine that nature has not made this rich landed country for an agrivation. In the third place, I have commenced, with a well arranged, powerful outfit. If I don't succeed, it will be the first thing of any importance that I began and didn't finish. . . . You, with your conclusions on the geological structure of this portion of the state, would expect to descend two or three thousand feet to get through. Well, I have no objections to going that deep. I must work at something, and I had as soon be employed boring a hole in the ground, as anything else. . . . You would "expect to descend 2 or 3 thousand feet through the cretaceous, below the tertiary," but would it not answer for me to bring up water from the *joint* between the tertiary and cretaceous formation? Or suppose, just for the amusement it would afford me, that I should violate the Geological theory, and force up water from one of the seams in the tertiary formation itself. Or, suppose for instance, as I have sometimes seen, that contrary to all theory and geological expectation, I should drop my auger into a river rushing through the middle of the Tertiary and find myself flooded with too much of a good thing. Geology is a nice science and I like it well, but I shall bore on at my well, theory to the contrary notwithstanding; because I have begun it.

The onset of the Civil War ended this unsuccessful experiment in November, 1860. Lincecum's personal interest in the study of geology was stimulated by his friendship with Samuel Botsford Buckley, a botanist and geologist originally from New York, whom he first met in January, 1860. Buckley was then on a walking trip that began in Philadelphia, traversing the continent to study forest trees and estimate their numbers. He soon decided to remain in Texas, and in October, 1860, he joined the staff of State Geologist B. F. Shumard as an assistant.[8] One month later Shumard was abruptly fired by Governor Sam Houston and replaced by Buckley. In the controversy that followed, Lincecum supported Buckley, and they remained close friends, despite Buckley's Union sympathies during the Civil War. Buckley recognized the value of Lincecum's accumulated knowledge and appreciated the freedom with which it was shared.

The State Geological Survey was suspended in 1861, and Buckley left the state. When it was reinstituted on October 30, 1866, Buckley returned to Texas as the state geologist under Governor James Webb Throckmorton.[9] In the spring of 1867, when Lincecum undertook his three months' exploration of nineteen Texas counties,[10] Buckley obtained a leave from the governor to accompany his old friend part of the way, in search of resources for economic development. Subsequently Lincecum submitted an extensive report to Governor Throckmorton (June 15, 1867)[11] commenting on such things as the evidence of coal, iron, marble, and petroleum as well as precious metals, the location of potential sources for water-powered industry, and even medicinal (sulphur) springs. As an outgrowth of this expedition, Lincecum had four articles published in the *Texas Almanac* of 1868 (pp. 85–91), as follows: "Gypsum in Texas," "Texas Marble," "Medicated Waters of Texas," and "The Water Power of Texas." These are reprinted in appendix 4.

Lincecum did not take his letter press along on this three-month expedition, undertaken to help him cope with his grief over the death of his beloved Sarah (on February 2, 1867), but the few letters we have include some interesting observations. To Spencer Baird he wrote from "Camp at the falls of the Brazos, Texas" (April 10, 1867): "I have made some collections of fossil, freshwater and dryland shells; from each stream I have crossed I take a suit of its shells, number them, and record them and the numbers where they came from. I have also made some collections of insects, but the weather has been too wet and cold for me to make much headway among the insects, yet I have captured some in nearly all the orders; most in Coleoptera, as a greater number of them survive the winter." On August 2, 1867, after his return to Long Point, Lincecum sent to the Smithsonian a box containing sixty-three specimens from the various counties he had visited. The accession file indicates that the collection was received August 7, 1867.

Still on the road, he wrote his daughter Mary Matson (May 2, 1867) from Mormon Mills[12] P.O., Burnet County:

I am very well, and though I have made but little headway in collecting Entomological specimens, I have stood the fatigue of travel and camp duties as well as any of the company. It has been a remarkable unfavorable spring. We have so far seen but four days that were pleasant and suitable for my business.

We have visited Austin, Round Rock, Cameron, Port Sullivan, Falls of the Brazos, Belton, Hawkins on Cow House Creek, Owl Creek, Nolands Creek, Lampasis river, Leon River, Lampasis springs, headwaters San Gabriel, head of Hamiltons Creek, Town of Burnet, and are now 10 mile[s] below at the Morman Mills, on Hamiltons Creek. We have had plenty of fish and fowl, and our horses are in better condition than when we started. No occurrence has transpired with us worth recording except perhaps a thing that turned up on the head of Nolands Creek, a few hours after we had left our camp. It was a great place for fish, and we had been there catching and cooking fish two days, and we lazily left our camp about 10 o'clock A.M. on my birthday [April 22]. On the evening of the same day, three Indians chased a man away from near our camping place, who very narrowly escaped them. This we were told by a traveller who overtook us two days after the occurrence. People everywhere warned us about the indians and more especially on Nolands creek.

In a letter to Joseph Leidy (June 30, 1867), he commented on the expedition: "I have a few days since, returned from my trip of observation. I remained out but 90 days until I had collected as much as I could carry. . . . I was compelled to leave the most interesting portions of my discoveries. I however brought home a right pretty collection nevertheless. Botany, Entomology, Geology, dry land and fresh water shells. My collection of Marine fossils and salamanders are valuable." In a letter to his friend Elias Durand (August 30, 1867), he waxed more poetic in describing the excursion:

. . . I found some grand scenery, some striking new feature in the split rocks, the beautiful unhacked forest trees, the pure untouched maiden botany of smaller types, and the glorious widespread carpet of uncropped grasses and nutritious herbage that clothed the gullies and up the mountain sides to their very summits as a full flowing garment. Twas at these times and in these cases that I thought of you, my dear D., and I imagined to myself the amount of pleasure it would afford me to look upon the workings of that bland old countenance and hear your remarks while you gazed delighted upon those vast, precious plats of untrodden nature. . . . How buoyant your good old, nature-loving heart must have floundered in its casement had you been by my side, on the brow of one of those cedar capped mountains, from whince, spread out before the smitten senses might be seen one of those indiscribable pictures, vast in its proportions, extending still, untill its dim extent eluded the eagerly enquiring sight.

For many years Gideon and Sarah made annual expeditions to the Gulf Coast of Texas each winter, and these allowed him to pursue his interest in geology. In a letter to A. G. Lane (October 7, 1863) Lincecum commented:

We make our trips to the Coasts during november and december remaining there till nearly Christmas. These are also joyous, healthy, intellectual excursions. I could consume a lifetime in the investigation of the wonderful and endlessly varied secret creations that are rolled to the fretted beach by the ceaseless throbbings of the heart of old OCEAN. Why really, it seems to me, to be the matrix of creation, and that all our dry land animals and vegetables have crept from beneath its waves. We know that all things from the chrystal, to the highest forms of organic life, first floated in water. The material of which the topmost bud, on the tallest tree is composed, was floated there by water.

Chapter 6

Texas Weather

The smell of the gray and sullen smoky norther always puts me in mind
of the odor that occurs while washing a dirty gun, especially if it is with
warm water that I am performing that filthy job.

Gideon Lincecum

The atmosphere above the soil and rocks of Texas interested Lincecum fully as much as the land itself. On his first visit to the Mexican territory of Texas in 1835, he had his initial confrontation with the striking weather phenomenon known as "the Texas norther." In his journal entry for February 26, 1835, he wrote:

> Lay by today on account of the severity of the cold. It was hailing, snowing & sleeting greater part of the day, wind from the north, very cold indeed. The weather in this country is the most variable that I have ever experienced in any country. Some days is so warm that you cannot keep your coat on, and perhaps a change will take place in three hours that will render you uncomfortable under all the clothes you can pack on. There appears to be a continual sesaw between the north and south winds, that is it has been so on our route ever since we crossed red river, having according to our recollection, not observed the wind to blow from any other direction than from the north or south, and what is most singular is, that you seldom experience a calm.[1]

In the years that followed his settlement in Texas in 1848, Lincecum continued to make references to the sudden northers, which seemed to impress him even more than Texas' legendary heat and dry spells.

Indeed, some of his most imaginative writing is connected with the Texas norther. To Professor Caleb G. Forshey of the Texas Military Institute in Rutersville, Texas, Lincecum sent the following report on December 5, 1859:

At 8 h. 5 m. P.M. last Thursday; Thermometer 73. Barmtr. 29.9. direction of the wind S.—barely perceptible; and while I was absorbed in making observations, by candle light, on the mechanical contrivances of a tolerably fair specimen of the ingenious tribe of animals, known and placed by the immortal Linnaeus as Arachnida—I mean the tribe not the individual—who under my observations, was so industriously employed, spinning and weaving his net in anticipation of fat things for supper that night. He had suspended the frame work of his device, between two widely diverging leaves of a large cactus—which flourishes on a little artificial stony mound in the immediate vicinity of the doorway of my shanty. [He] had finished tying all around the radiating threads to his outside leaves; was now at the center, looping in the meshes, and he had progress[ed] three rounds and a half, when he suddenly suspended operation. I began to think that he was fatigued, but at that moment, [I] felt the first gentle, but chilly breath of the norther. Then it was calm again, and he quietly resumed the work of measuring and tying the meshes of his net as before.—Soon there came a fiercer blow; he hesitated not another moment, but dropping all, and clinging to the thread [with] which he was, an instant before, so busily and so artistically constructing his distinctive trap—cast off, and letting himself safely down, crawled into concealment amongst the tangled oak leaves, which had accumulated on the foot of the cactus.

Poor fellow, thought I, you have missed your supper, but you may console yourself, for it won't be long till you will be insensible to the pinching demands of a hungry stomach. My mind then taking a wider range, pensively inquires, how many short-sighted beings in this *providence protected* world of organized life, will be, like the spider, overtaken this night, in the midst of fondly cherished anticipations, equally unprepared to meet the fierceness of the pending Norther, [having] dropped the thread of their respective vocations, never to be resumed again?

The heavens were soon overcast with cloud, and the wind was quite gusty. This time, I could detect no peculiar odor. One hour after, [at] 9 P.M. the time for making and recording my noc[turnal] observations. Going the round of my instruments, I found the anemometers— I have two—marking direction of the wind to due N. Force of velocity, 40 miles—Amt. of cloud, 6—thundering and sprinkling rain which in the rain gauge by morning amounted to five eighths of an inch. Thermometer, 53. & rapidly sinking. Barometer, 29.8. Dec. 2 [next day], 7 A.M.: Ozone 4. Thermtr. 30. Bartr. 29.9 Wind N. force 4 velocity 40 m. Amt of cloud 9, and homony hail [i.e., pellets the size of hominy] falling fast.

By next evening the temperature stood at 12°.

During the heat of the next summer, Lincecum's thoughts turned, perhaps longingly, to the winter northers, and he wrote imaginatively to Forshey on July 1, 1860:

As for me, after working myself into the clairvoyant state, thoroughly inflating, with pure theoretic god, the largest balloon of my imagination, I made my ascension from a snow capped peak in the rocky mountains N[orth], riding merrily southward on the steady sweep of the dry fogless norther, overlapping the swift flying, broken clouds which were passing in an opposite direction beneath my feet. Whilst thus smoothly, steadily floating, southward went my aerial ship; I was rejoicing at the ease and speed with which I was overriding the darkened stratum below, when suddenly, and right ahead, hove in sight, a belt of black, fringed with scintillations of vermicular light, stretching wide athwart my path—onward, and fiercely sped my *norther* driven car; soon, very soon, I was in dreadful proximity of a frightful gulf; I could perceive that the smoothly gliding *norther*, upon which I had been so felicitously and fearlessly sailing, here bent abruptly downwards, forming a precipitous cataract of 10,000 feet descent; against the solid impenetrable front of this *norther* cataract, fiercely beat the frustrated and underrun South wind; here abruptly checked, the drifting mists and humid sea clouds rapidly accumulate; cloud jams on cloud and the storm god bellows; burning flashes of liberated electricity cleave the gethering blackness; sulphurous fumes, accompanied with jarring thunders, greet the frightened senses; urgent south winds pressing up from behind, squeeze from the black, uptending clouds a moderate shower of needed moisture. At this moment of wild commotion—this din and uproar—my fragile balloon, reeling and swaying at the forward moves. Now on the brink, she tilts adown the awful cataract, makes the frightful plunge, still whirling downward through the dreadful strife of conflicting elements, its inverted form, ignited by electric fires, with dire explosions bursts; and with its smoking fragments, the proud, aerial navigator—your humble servant—is unceremoniously dashed, and rudely tumbled, half frozen, on the grass clad plane below. And now the fierce *norther*, sweeping the prairie, underruns the moisture bearing south wind, from which, with thirsty gorge it soon drinks up the entire burthen of the S. wind. But in its eager fierceness, the norther sinks the temperature below the freezing point and before it has time to appropriate the water obtain[ed] so rudely from South wind, a vast amount of globules are necessarily frozen; being highly electrized, these congealed globules are attracted by the twigs of trees, and weeds, and grass, and circling spiders webs, clothing them, when the bright sun shines out, in glittering gems and gorgeous habiliments, displaying in beautiful fantastic figures, all the tints of the seven fold threads of light. Sometimes—Yes, I have often seen it raining upwards, when it was trying to rain downwards. Saw it today.

In a follow-up letter on August 17, 1860, Lincecum indicated that Forshey evidently did not know how to take his account of the norther:

Burlesque you suggest. Not a bit of it. On the contrary I am truly pleased at my success, by the little bungling description of my imaginary aerial voyage, in proving to your mind that I understand your theory of the peculiar action of the Norther. I had often observed the violent ground sweeping action of the sudden Norther, but had not conceived the idea, until I read your paper on the subject, that it is caused by the pressure of the superencumbent stratum of South wind under which the norther, wedge-like, is fiercely and forcably driving its unceremonious and boistrious course. The sudden Norther, bursting upon us as it often does—like an explosion, indicates, as I think, that the cataract is not very distant; but when it comes by, at first, a very gentle breath of cool air, after a few minutes a little stronger, then again still stronger, like it did the night I was watching the spider weave his net, it indicates that the place of the downward plunge is far off; and this, I think is somewhat proven by the constant fact that the insidious [i.e., stealthy rather than sudden] norther does not sweep the earths surface with so much trash-lifting violence as the sudden [one]; nor is it attend[ed] with so much fog and the peculiar gun washing smell of the sudden Norther.

Although his mentor Erasmus Darwin would have appreciated this romantic description, Lincecum did more than use his imagination to speculate and theorize concerning Texas weather. One letter to Navy Lt. Matthew Fontaine Maury, August 4, 1859, suggests that he had already been keeping weather observations for ten years, which means that he began doing so very soon after settling in Texas. For example, he collected data about the arrival of one norther, as reported in a letter to Forshey, November 17, 1859:

The "terrible Norther" you mention—12th inst.[Nov.] struck us at the same moment—

[12th]   7 A.M.: Thermometer 70[°], Bartr 29.07;
             2 P.M.: Thermtr. 34°; 10 P.M.: 28°;
13th   7 A.M. 20°—2 P.M. 46°—10 P.M. 33[°]
14th   7 A.M. 20° 2 P.M. 58° Wind S.W.

The norther which you have named as occurring on 24 Decr. 1855 struck Mount Olympus [Lincecum's home] at 3h. 53m. A.M. The smell of the gray and the sudden smoky norther always puts me in mind of the odor that occurs, while washing a dirty gun, especially, if it is with warm water that I am performing that filthy job.

Twelve years later, writing in his diary on September 10, 1867, Lincecum observed: "A norther came this morning making all the forenoon pretty cool; people all put on Thicker coats. It was a smoky norther, and although the clouds that followed it were black and threatening we did not expect rain. The smoky northers as a general thing do not indicate rain. I have seen several dry winters since I resided in Texas and the northers were all smoky."

Lincecum was very interested in the climate history that trees could provide. He was also aware of the research on tree rings of Jacob Kuechler that was first published in 1859.[2] In a letter to G. J. Durham, September 1, 1859, Lincecum commented on his comparing the work of Kuechler on climate cycles in Texas (based on trees in the Fredericksburg area) with his own analysis based on trees in the Long Point, Texas, area. He states:

Three years ago, a number of our citizens, on account of the severity of the prevailing drought, had become restless and would, very probably, have left the country, had they not been taken to the silvan record, and instructed in the manner of reading the sacred, truthful books. Large old trees were cut down, sections sawed off, made smoothe, and the mycroscope resorted to, when the "year rings" were too small to be correctly observed by the naked eye. The trees which were selected for the purpose, were the same species of Oak (*Quercus obtusiloba*) as those experimented with by Mr. Kuechler, and as in his experiments, were taken from the elevated, dry lands. Our largest trees run the record back 141 years, and it is a remarkable fact that the results of this reading of the "year rings" of the oaks of our country are almost precisely the same as that had at Fredericksburg. . . . That the present drought is an exception to any that has occurred within the last 141 years, I may add a scrap of testimony, resulting from my own observations, which will, in my estimation, strength this opinion considerably.

The pin oak (*Quercus palustris*)[3] delights in low moist lands, is a large useful tree, is found in groves along the narrow bottoms of small creeks, and sometimes in wider bottoms. During 1856, this species of oak, in our country, almost entirely died out. Old trees and young saplings all shared the same fate. Some of the largest that I examined contained 143 year rings, and as the old and the young have alike all died together, the conclusion that the present drought is an exception to any that happened within the period of time recorded by these pin Oaks is, it seems to me, quite a fair one.

Lincecum read Matthew F. Maury's *Physical Geography of the Sea* (1855) and volunteered his services as a weather observer to Maury in a letter dated August 4, 1859: "I reside in middle Texas, betwixt the Brazos and Colorado Rivers, 100 miles from the Gulf Coast. Lat. 30° 20′ N. If it will be of any service to you in carrying out your plans in the very laudable and extremely useful work, in which you are so energetically engaged, you may forward to my address, Admiral Fitz Roy's circular and a set of blank forms for keeping account of the wind and weather." Again, on September 30, 1859, Lincecum wrote Maury and enclosed his weather data for the month of September. He wrote, "A word of instruction on that subject may not be amiss; and as I am anxious to aid in this business as far as my ability and instruments will permit, any instructions from you, relative to its correct prosecution, will be thankfully received and attended to."

In his letter to Forshey dated November 17, 1859, Lincecum reported:

I have no guide for keeping a register of the wind & weather but the tables sent me from the National Observatory and my instruments. 2 Thermometers, a Barometer, Ozonometer,[4] and 2 wind machines, which I constructed myself. One of which weighs the force of the wind; the other, very correctly measures its velocity; and as far as I am concerned they are both original. The N.W. wind at 3 P.M. today passed at the rate of 40 miles an hour. At 10 in the forenoon at the rate of 23 1/2 miles an hour. It was then from the S.

I have this moment—10 P.M.—returned from an observation on

the Anemometers, they give the direction of the wind N.W., force 3, velocity 2 & gustly. Thermometer 47°, Barmtr. 29.09. Clouds, none; mercury sinking.

Maury told Lincecum to send his monthly reports to Admiral Fitzroy, Board of Trade, London, England. On November 1, 1859, Lincecum sent the October report to Fitzroy and enclosed a short note: "My instruments are new, and as yet, not entirely regulated. Shall be able to make my observations more exact soon." Receiving no instructions or acknowledgment from the National Observatory, Lincecum eventually stopped sending reports. He began keeping records for the Smithsonian and continued doing so until 1867, when he passed the weather observations to M. Rutherford, the postmaster. In a letter to Prof. Joseph Henry, Smithsonian Institution (June 20, 1867), Lincecum states that Rutherford "is a sober and very correct business man. He has commenced recording the weather indications three times a day, and will, I have no doubt, make satisfactory reports."

It was common to see Lincecum's reports in the *Galveston News* as well. For instance, in a letter addressed to "Editors, News, Galveston, Texas," dated December 1, 1860, he wrote:

The following indications, copied from my meteorological journal, for September, October, and November 1860, is respectfully submitted:

During the month of September, the average temperature at this point, at sunrise was 70 - Fah; at 2 P.M. 89.01. The highest point reached, 98. Whole amount of rain, 2 7/8 inches.

Average temperature for October, at sunrise, 56.07. The mercury stood as low as 45. on 10th and 20th. Average temperature at 2 P.M., 77.01; highest point reached 94. There were, during the month, five showers of rain, amounting together, to 2 1/2 inches. First flock of wild geese passed south on the 8th.

For November the average temperature at sunrise was 50.08; on the 24th the mercury stood as low as 30, at sunrise. Average at 2 P.M., 66.01; highest point 84. Total amount of rain, 4 1/2 inches. First and only ice occured on the 24th. First flock of swans passed south on the 21st at 6 1/2, A.M. The month closed with the grasses all green, and the entire genus of ants were deepening their habitations. Do they know anything about the meteorological indications?

Knowing only too well the vagaries of Texas weather, Lincecum was too wise to trust the prediction of even his "brother emmets" without some reservations. Writing on September 22, 1866, to Elias Durand, he reported: "I recollect one year here, on the 9th of January, we had had no frost, and for dinner we had snap butter beans, squashes, new potatoes, tomatoes, and there were peaches that had grown from blooms that occurred in November, as large as the ends of my fingers, and I was begining to hope that I should live to see some bienial peaches, but the norther came at 9 P.M., and at 6 A.M. the next morning the ice was a foot thick, to the great damage of my cucumbers, squashes, peaches &c."

Texas folk wisdom holds that "only a fool or a newcomer (sometimes translated 'Damnyankee') predicts weather in Texas."[5] Lincecum was no fool and by 1860 hardly a newcomer, so he contented himself with weather reporting rather than weather forecasting and let nature have its way. In Texas, nature's way meant heat and drought as well as "Northers," and sometimes even Lincecum came close to complaining about the former. Writing to his nephew John Lincecum on July 10, 1860, he said:

> I write a letter to somebody every night—some letters takes me two nights—but as the weather is so intolerably hot now, and has been for 2 months, I cannot write long letters. The hot weather, or something else, maybe its old age—but I have no reasons to complain of that yet awhile—affects my nerves so much that it is with difficulty sometimes that I can write at all. I have never experienced such warm weather before in my life, in any country or season. It is now 9 P.M. and the thermometer is standing at 100°.
>
> Last Saturday, at 2 P.M. it stood at 108°—the average at 2 P.M. for the last 2 or 3 weeks, has been 103°, and at sunrise for the same time 74° and very rarely a cloud to be seen in any direction. We have had no rain since the 22nd of April; the consequence is that there is not corn enough made in the country to bread the inhabitants; as for pork and feeding horses, thats out of the question. What little corn we have made is now perfectly dry, and we intend to gather it next week. The last two days we have been driving our stock off to water; and the nearest is five miles off. The grass on the prairie is all dead, and as perfectly dry, roots and all, as a tinder box. The wells this year have all been deepened 10 feet, and water is scarce at that. I have been forced to suspend my well boring operations for want of water to run the engine.[6] The prospects are discouraging, and many people are seriously frightened and talk of leaving the country. But the drouth seems to be universal, and I know not where they can go to better themselves.

It would take more than heat, drouth, or northers to drive Lincecum from his beloved Mount Olympus, but in 1868 Yankee Reconstruction drove him not only from Long Point but from Texas as well.

Chapter 7

From Texas to Mexico

*Since I lost my Dear old companion, with whom I had spent 53
happy years, I can find nothing that can fill the vast chasm it has produced
in my enjoyments—for she always went with me in my various
excursions. Not being able to remain long at a time at home,
I have become like a wild turkey gobbler; I do not like to fly
up to roost twice at the same place. But wander whithersoever I may,
I shall not forget the subjects of natural history.*

Gideon Lincecum

By 1867 a number of disgruntled, disgusted, and defiant Texans had emigrated to Mexico to escape the yoke of Yankee Reconstruction, and others were making plans to follow their example. One such colony in the valley of the Río Tuxpan in Veracruz State was being promoted by John Henry Brown, a Texas journalist and Civil War veteran.[1] Brown's glowing accounts of tropical Mexico captured Lincecum's attention, especially after his world was shaken by the loss of his beloved wife, Sarah, on the second of February, leaving him profoundly empty and lonely. He expressed his feelings movingly in his response (March 3, 1867) to a word of sympathy from his good friend Elias Durand:

Thou art my brother! for who but a brother could have touched the subject of my greatest distress so tenderly? Yes tis hard indeed, after so protracted a companionship, with one of no ordinary endowment for goodness and all the thoughtful attributes of kindest attention and affections, to be thus torn asunder; and that separation, is it forever? I am a man, and can think correctly on such subjects; but with all the power I possess, I can not look calmly upon the turned plate at the table, nor upon the old empty chair on the left hand side of the fireplace where she so industriously worked embroidering little presents for her very numerous Grand children so long. To remain here, will not do for me now. Everything I see, minds me of my lost one ____ ____ ____ [sic]. I am of no use here nohow. I look upon it now as a good thing that while she yet lived, I had made arrangements to spend most of my time in the wild uninhabited regions, where she hoped to be with me, and encouraged me to go, at all events even, should she fail. I shall most certainly do so, and ten days will see me harnessed up, and on the route. I have nothing now to veer my course, nor to call me back. My children all have homes of their own and are doing well (nine families). They do not need anything further from me. Therefore, I see no impropriety, nor can there be any, in wandering in the hills and forests, and devoting my time and means to the only thing that gives me pleasure now.

Although Lincecum busied himself with scientific and pleasurable excursions to the hills of West Texas and to the shores of the Gulf, his bereaved soul still felt empty. To his youngest daughter, Sarah, living near Houston, he wrote on February 8, 1868: "A great change has taken place. The family ties—the golden bowl—has been broken, and the occupants of a long loved and highly cherished homestead have been scattered. The fields and houses are filled with strangers, and I, as the once acknowledged head of that flourishing family, stand alone and howl like a lost dog, in utter amazement at my irremedial loneliness ____ ____ ____ ____ [sic]."

This loneliness combined with Lincecum's extreme disgust with Yankee Reconstruction and his restless, adventurous nature that had led him earlier from one frontier to another, from Georgia to Texas, now prodded him to pull up stakes once more and leave Texas and the United States behind. His bitterness over Yankee rule was magnified by the removal of his new friend, James W. Throckmorton, from the governor's chair by General Phil Sheridan on July 30, 1867.[2] Finding himself ineligible for any elective office or official position and his ex-slaves enfranchised, Lincecum replied to his scientific friend, Alpheus Packard of Salem, Massachusetts, on November 20, 1867: "You ask me why so many of our citizens are going to Mexico? I answer to get away from Yankee oppression and nigger equality."

Added to his loneliness and disgust was the restlessness he felt, like his father before him, as settlement increased around him. To his friend S. B. Buckley, who reported "the probability of 200 Virginia immigrants" coming to Texas, he replied December 12, 1867:

To desire that the country shall be filled up, the grass all killed with the plow; our cows all starved out of existance, and our roads all turned into the ravines and gullies, is an unholy desire, and can have its origin only in the hope, that the greater the number the more fools, and that the chances for making money by fraud and misrepresentation will be increased by the increase of numbers. The conclusion is that everybody is willing to try his luck at the game of fraud. —I may say that I have arrived at the end of my road now!

And to Dr. Engelmann in St. Louis, Lincecum wrote (November 25, 1867): "Now society, villianous, *civilized* society, has gradually thickened around me, until there is no little, sacred, untramped nook to which I might retire and hide myself from the unholy gaze of intermedling civilization. Hence I go to the wilds of the mountain forests of Tuxpan." In a similar vein, his diary entry for July 6, 1867, reads: "Collected some butterflies and grasshoppers. I also put my article on the marble of Texas into ink [revising a pencil draft]. It is, I think, a pretty respectable paper. My mind however is becoming gradually more indolent and indisposed to study & to grapple with serious or heavy propositions. It requires exciting subjects to stimulate it into action; it would become useless if I lie up in dull society. I must wander over natures rugged fields."

Although Lincecum was eager to go to this new frontier and thought his age, which by this time was 74, should be no barrier, his family and friends thought otherwise and sought to dissuade him. Lincecum could boast how well he had withstood the three-month expedition to West Texas, but an excursion in August, 1867, to explore nature's treasures in deep East Texas almost proved the undoing of his Tuxpan plans and of himself, as he reported to Elias Durand on August 30, 1867:

Your very kind letter of 24th July ult. was on my table when I returned 17th inst., from a tour of observation in Eastern Texas. I returned quite sick, and have not [been] able to pay any attention to the basketful of letters that have been awaiting me, until yesterday. You must not lay this sickness to the hot weather and the fatigues of mountain travel. The country into which I went is flat and marshy, full of slime and mud with its accompanying barrels of vocal music, making night glorious with their hidious notes from the extremely hoarse, earth jarring voice of the great alligator, down through the quite respectibly sounding Bullfrogs, and many grades of lesser fry to the barely audible "neep,neep," little singers in human shape, not longer than a ordinary house cricket. It had rained a great deal, the whole earth was afloat with water, and having no boat, I could only stand and listen to the music. The yellow fever was also there, and as I could affect nothing, I thought it prudent to turn away from the songs, prayer meet-ings, muskitoes, and other wonders of that unexplored region with-out attempting anything further on my part. But [Sioux] Doran was sick; few hours more Sarah [Lincecum's daughter, Doran's wife] gave up, then the negro boy we had along riding our loose horse, and by the time I had got Doran by vomiting him a whole day safely on his feet again, I felt sick myself. Sarah was very sick; and the negro lay and slept on the horse, for we did not cease to travel. Doran kept the ambulance[3] rolling, while Sarah and I continued to take the emetic, hauling out an occasional dose to the benumbed and sleeping african. It was that preparation of Lobelia, called Antispasmodic Tincture, by Botanic doctors, which requiring only water as a vehicle, we could conveniently take it, which we did, continuing the vomiting through the day, till sunset, when we were so much relieved, that we hoped, like Doran, we had thrown off the yellow fever complaint, for it was of the yellow jack type;[4] not so however, it came again and again, baffling all efforts to checking it. We were five days on the road, vom-iting away the fever; we [then] turned about and reached home, camp-ing out all the time. People would not offer us, had [we] tried, [nor] permitted us to stop in their houses, for they think of us as travelers infested with some contagious sort of fever . . . from the East. But the camp in such cases is far preferable to sleeping in their crowded and illy ventilated cabbins and decaying houses. At the camp, every breath you inhale is a new one, which I consider a great advantage, at a time when the vital forces are almost equally matched in a conflict with any important form of disease. As to my own [self], I have no doubt, had the journey been two or three days longer that I would have reached its termination clear of disease. As it was, I reached home with some fever and dysentry. Crowds of visiting friends, hordes of my numerous broods of children, including my three doctor sons,[5] talking incessantly an unknown tongue, soon breathed what little air my room contained about forty times over, and I [became] worse. They stood around looking wise, and remarking [how] on getting home, bad cases always got worse. [I] saw the grounds of their superstition and cried out murder, Oh! take me outdoors and lay me under some shady tree where I can get one pure breath of air. But they looked at each other and nodding significantly, said his mind is giving way. My doctor sons succeeded in clearing the room when nearly too late, and I wallowed in fever and science, tipped off with a few fashionable remedies, five days longer; when the fever left me, thin and flat and with but little of the vital forces to go upon. And this was brought on, not by the weakness of old age and fatigue, but from going to the wrong place, not into the mountains and valleys, but into a region of mud and slime, the home of all the reptile races, from the slimy ground puppy to the crockadile.

Far from being discouraged by this experience, Lincecum was prepared to attempt an even more ambitious undertaking than settling in Tuxpan. In the same August letter to Durand (August 30, 1867), Lincecum added:

> In regard to leaving the U. S. I changed my programme a little. I think it quite probable that the U. S. intends to possess themselves of Mexico shortly, and then should I go to Tuxpan I would have to move again, so I have discarded that plan, and have determined to go direct to British Honduras. As soon as I can write with a little more ease, I shall address a letter to my old friend and correspondent—until the war broke it up—Sir Charles Darwin, informing him of my removal to [British] Honduras, with a proposition for us to resume the investigation of the Ant genus, including the Honduras ants and all subjects found there, belonging to natural history, and to his favorite theme of the "Origin of Species." As he is a prompt correspondent, I shall expect to meet his letter at Balize, and receive encouragement that will enable me to go into the mountain forests at once. What do you think now?—I am not going begging, I have the means to support me through all this last named programme.

There is no evidence that Lincecum actually renewed his correspondence with the great Darwin, who no doubt would have been astonished or amused by Lincecum's uninhibited proposal. In any case Lincecum soon dropped the British Honduras option and focused his sights on Tuxpan. To his medical friend, Dr. H. C. Parker of Houston, Lincecum wrote on December 3, 1867:

> You think I am too old to emigrate to distant countries. I do not perhaps understand that. I was thinking that I am too old to remain here much longer. I am however, strong and active, without a defect or blemish about me, and I am tired of remaining in a country where there are no new subjects for investigation. I have this year explored 24 counties of our state, in addition to what I had been doing for eight or ten years before. There are but few new things, either in botany, geology, Entomology, and several other minor branches of natural history for me to find in Texas now. My mind and spirit are as live and vigorous as ever, and to think of staying here, in downtrodden, africanizing Texas is too horrible for me to entertain the thought even for one moment. I must get away from the U. S., the Negroes and the more disgusting radicals. . . .
>
> Prof. Ruter[6] had informed me of your intention to visit Tuxpan,

and I was highly elated at the prospect of exploring that lovely valley in company with you. From a long passed experience, I know that in the wildest forests, we could spend our time profitably and live comfortably. The extensive unexplored fields and forests in the Tuxpan valley and the adjacent mountains. No scientific exploration has ever been made in that country, on which account I anticipate a grand intellectual feast.

As the year drew to a close, Lincecum caved in to his children's entreaties and suspended his plans to journey to Tuxpan. Almost disconsolately, he wrote to Spencer Baird on December 11, 1867:

When I was a very young man, I read Dr. Franklins works. He advised early marriage, and that advice agreeing with unchecked and misdirected amativeness, it was an easy matter for me to fall in with the old sages directions. Accordingly I sought out a companion and was engineering the matrimonial machinery before I was 21 years of age. The result is ten families of grown up men and women, with their children, numbering together 61. I do not repine, or regret anything about it, but I cannot avoid the recollection of the fact, that in rearing this numerous brood—who average only from ordinary to good middling—I lost 38 years of a life that could have been better employed; for the world is as full as it can hold of precisely the same sort of folks, and there was no use in adding my brood to the already overdone business. The demonstration of the fallacy of the Franklin theory lies in the fact that there are ten real, boni fide families of the same old sort of people, and notwithstanding that they clogged my action 38 years, and that I have recently divided my entire landed possessions amongst them, they are not satisfied with what they have already done by way of checking my progress. But they must, now I have got all ready for my contemplated excursion of observation, get around me and cry and grieve, and cut up shines, declaring that it never will do for me to leave them. That I must not go away, &c. &c. They have confused me. They have shaken my heretofore strong unwavering resolution, and unfitted me for future progress. I shall be compelled to abandon my Mexican exploring trip. This will disappoint two Companies of emigrants who will form a junction, this day, on the road to Mexico, 30 miles from Long Point and will expect me to join them.

And now I may consider myself as being laid up, to remain in waiting for the final close. There will be no use in that. I have examined everything belonging to Texas, preserved everything I had the means of doing it with. I could not bear to preserve snakes, lizards, salamanders, fish, &c. in $8 alcohol, [even] if I had been able to obtain suitable vessels to put them in. Dr. Cope furnished me last spring with a list of specimens (of snakes, alligators, catfish, &c.), sufficient for a barrel or two of Alcohol. For reasons which it does not require that I should name here, I did not make the collections. I have not yet thoroughly discovered the class, or rather, species of the genus homo, to which those Philadelphia quakers belong. I shall study them more carefully, and if I find them worth preserving, I can furnish the alcohol myself to put them up in.

It is not likely that I can find any birds or other specimens belonging to the Natural sciences in Texas now that would interest you, or add anything to natural history. If, however, I should at any time become inspired with sufficient courage to make excursions in the forests and along the water courses, and should find anything interesting, I will send it to the Smithsonian.

To assuage his great disappointment and to prove his hardiness to his protective children, Lincecum proceeded to undertake an expedition to the Gulf Coast. Once on the way his spirits soared, and he wrote a glowing report of his safari to his daughter Cassandra Durham (January 19, 1868), emphasizing how durable he was, capable of undertaking any expedition, even one to Tuxpan:

Tomorrow will be two weeks since I returned from the Gulf coast. Having made all ready for my Tuxpan journey, and failing to collect sufficient funds for the trip, I could not go. The disappointment was pretty hard to put up with; and having my outfit all complete and ready, in the delirium of my disappointment I packed up my nice, substantial fixtures and set out for the Coasts. The weather, though threatening, continued very pleasant, and I, traveling slowly, shooting quails, squirrels, ducks and any eatable small game, gradually approached the salt water. Saw more deer, geese, ducks and snipes, west of the Brazos, in Fort Bend County, than in all the balance of the excursion. We crossed the Brazos and at Stafford's Point, 12 miles below Richmond, got directions from a Negro, who pointed the course to the head of chockolate bayou.[7] There is no road, and we struck off over the wide expanse of prairie. We had not travelled many hours until we were entirely out of sight of land, nothing to be seen in any direction but the shoreless ocean of grass. Fowler,[8] who was with me, becoming uneasy, thought we should have provided a mariners compass for the exigences of such a perelous predicament. It was not very long, however, until far away in the dim Eastern horizon, we discovered the tops of some trees, in the proper direction, according to the Negro, for the head of chockolate bayou. Jimmy could not imagine by what faculty it was that I had travelled so far on a direct

line, when there were no landmarks to go by. But you must recollect that Jimmy Fowler is not an American. That was not the only strange things he saw and heard during the expidition. It was night when we reached the bayou, and camping there at the little grove of timber, far out from any inhabitants, we had good fires, plenty to eat, and thick beds of dry, curly grass to sleep on. Had it not been for the frequent wolf concerts we might have done well, but Jimmy did not admire the music of the wild, grassy plains, and would keep talking, till George,[9] becoming impatient, declared that the wolves were less annoying, and their roudy uproar more to the purpose than Jimmy's howlings. Next morning was promising, it being the day before Christmas, and having passed safely through the dark, wolfy recesses of a prairie night, Jimmy was lively; and by sunrise [we] had all ready and pursued our journey down the bayou till 4 o'clock P.M., when we struck camp in a Yopon thicket near Mr. Willbourns,[10] a very religious man. For all that, he let us have what corn we needed at $1 a bushel. Next morning at break of day, I tuned up my old, black violin, which I had taken along for the purpose, and having no doorway to stand in, I stepped barefooted into a little gap in the Yopon thicket, and performed the Christmas tune[11] in good style, repeating it, as usual, three times. And that made 57 Christmas daybreaks I have done that, without missing a single Christmas. It is said, that "what man has done man can do again." That saying is not always true. In the case above recited, I may say, with much confidence, that to perform it is out of the power of any living man. Old Mr. Willbourn came to our camp to enquire for the meaning of that short round of daybreak music; when I informed him that I belonged to a new religious sect, and that to play that particular piece of music, three times over at daybreak, was one of our modes of worship. He thought it very strange, and left.

We got off early and by 10 A.M. came to an old but small and much dilapidated town called Liverpool,[12] on the west side of the bayou, which has here widened into a pretty little deep river, navigable for steam boats. 3/4 of a mile below Liverpool, and [at] his houses, which are very comfortable, on the bank of the bayou, I found my good friend, Samuel Adams[13] and family. They were all very glad to see me and made my arrival the excuse for a big egg nog, as it was a holy day and the family all Methodist. They brought it around in foamy-topped, large glasses; and so as to avoid giving the dear Christian friend oppertunity for any remarks, I took a teaspoon, which I found in the glass presented by a little girl; and dipping it full of the frothy, stinking stuff, gave a merry christmas to the pretty little, innocent creature. [I] turned off the filthy beverage, and returning the spoon, told the little girl that she might have the balance of it; at which she seemed quite rejoiced, and soon drank it up. She seemed to carry it well. From Adam's to chockolate or west bay, it is, by land, to the fishing place, 10 miles; and by water 25 miles. There is no road, and as I did not know where the fishing place was situated, Adams, his little son and George, went down in the Ambulance; and I, with Jimmy and an equally awkward Negro for oarsman, went down the bayou in a good-sized boat. We had strong head winds all day, which retarded our progress so much that we did not reach the fishing place by 4 or five miles. The bay had widened to 5 or 6 miles, and night setting in, I could see nothing but water; and not being acquainted with that portion of the Gulf coast, I concluded to go ashore and wait for the morning. We found some driftwood and passed the night pretty well. . . .

Although I arose half wet from my muddy berth, it did me no injury; and by the time the morning light was sufficient to distinguish objects a hundred yards, I was steering my course down the broad bay, and as it was calm, we seemed to make rapid headway. We were all looking out along the low flat shores, anxious to discover the campfire of our friends, when just at the moment that the glorious sun heaved his broad flaming disc above the glittering waters, right in his very face, as if thrown there by a daguerreotype, the Ambulance with its front curtains rolled up. [It] was very distinctly visible, horses and all, to our whole party. In that direction, from the flare of the great sun, no land was visible; just the carriage and harnassed horses seemed to hang in his very center. It was a most beautiful scene. A sweet, serene, clear morning, the transcendently glorious sun heaving up his broad flaming orb from the wide expanse of glistening waters, with a black ambulance and harnessed horses hanging in his center. [It] is a sight not only beautiful and wonderful, but one that will not be presented more than once in a lifetime. It lasted but two or three minutes, when the splendor of the scene changed; the Ambulance, now dwindled into a toy wagon, could be very plainly seen in the space beneath the uprising sun. It was distant about five miles; we rowed on and by 9 A.M. rejoiced our hungry, shivering friends (who had remained there all night, without fire or anything to eat) by giving them the needed articles—we having carried the luggage down in the boat, to favor the horses. We were all very happy now, and getting ready our fishing tackle; we fished and fried and eat, and fried and fished and eat; fat red fish and flounders, till we were all gorged and sick of fish.

All being cloyed with fish (there were no oysters, all having been killed last summer, by the bay being kept too long filled with fresh water), we caugh[t] and salted up a big boxful of redfish on saturday afternoon, (28th. December ult) and by daylight next morning set out for home. I took charge of the Ambulance on the return trip, and its a fortunate thing that I did so, for there came one of those cold thunderstorms with its flood of rain, when the party in the boat were exposed 2 hours to the merciless pelting of a drenching rain. We all got home safe, however, except George, who in a play with Adams' boy, fell out of the Ambulance and the hind wheel ran over his left leg and right arm and came very near breaking both. It did him no serious injury, and all parties arrived home just in time to meet the light norther which followed and kept us indoors 2 days. Next morning, 1st day of January, the weather was clearing up, and bidding adieu to our friends, we shaped our course over a pathless prairie to the head of Dicki[n]son's bayou, a tributary of Galveston bay. I thought I would have a fishing frolic at the mouth of this bayou, but on reflection, I could see no use in it. I did not want to eat any more fish, and giving that expidition out, I struck out over the interminable fields of grass, shaping my course for the City of Houston, where I expected to join Sarah and her consort. They had promised to remain until they heard from me, and the programme was that if I could find a suitable place, they were to join me there and we were to spend the remainder of the winter fishing and shooting. I found the place and rented a very comfortable house, where the tides ebbed and flowed every day, full of fish and all kinds of wild fowl and other game quite handy. Besides, it is a most delightful situation, with good water, sail boat, and good health. When I came to Houston, however, Sarah and Sioux had left on the cars for Columbus. They had dodged me and I came home.

Lincecum also reported his achievement to Elias Durand (January 29, 1868):

Being all ready, to show them that I am not the old decrepit invalid they wish me to consider myself, I packed up my good Ambulance with my Tuxpan preparation, and set right out for the Gulf Coast, distant 125 miles. I arrived there on Christmas day. Camping out every night, sleeping on the ground without spreading my tent, except when it rained. I slept on the salt marsh into which, during the long nights, I would sink deep enough for the salt slime to ooze through my blankets and wet me to the skin. During the day, at noon, I wallowed for an hour in the briny sea. Of nights, before bedtime, I waded in the salt water, shooting flounders (gen. Plateson)[14] with my bow and arrow till 9 P.M. While there and on the return route, encountered two severe northers. All this passed without any inconvenience, undue fatigue, or sickness of any kind, and I am in as good health now as any body. This good health and strength and power of endurance, is attributable, not so much to my good robust constitutional forces, as it is to a knowledge of the means to keep the said constitutional forces in the best possible trim—condition—a knowledge of what to do and what not to do, and to do it promptly.—And thus I have demonstrated to my friends and my progeny of 61 souls, that in reference to my physical inability, they are mistaken. This test experiment consumed a month or more of my time, during which I did not write any letters, and that is my excuse for not writing to you. Is it not a lawful excuse?

Lincecum did succeed in quieting the opposition to his removal to Tuxpan, and in June, 1868, together with his widowed daughter Leonora Campbell and her children, he sailed from Galveston for his new tropical and scientific frontier in Mexico. His last letter before leaving was to his sister Emily and her husband, D. B. Moore (June 4, 1868):

This is perhaps the last letter I shall ever write in the U. S. Certainly the last from Long Point. At 8 A.M. tomorrow, in the company of Leonora and her seven children, I shall set out on my journey for Tuxpan Valley, Mexico. We shall take the cars at Brenham, Saturday at 7 A.M.; the next morning (Sunday), we shall wake up in the city of Galveston, where we shall immediately go aboard of the Schooner *San Carlos*, and she will depart from that port Monday afternoon, for Tuxpan. The voyage is made in 5 or 6 days, in ordinary weather. It is a lovely country.

> We go to a far southern land,
> Where the coffee and sugar cane grow,
> Where the banana and Orange tree stand,
> With the pine apple close by the door.

It is eleven or twelve degrees south of this country, where the days and nights are always the same length. I shall live 10 or 15 years longer in that country than here, and find a great deal more amusing excitement amongst the new birds and flowers and butterflies.

We are all in good health and spirits; and, although I am 75 years and 43 days old at sunrise this morning, I am still robust and active. We have no interesting news in these diggins. I will write you from Tuxpan.

Chapter 8

Gideon in Mexico

What would I care for the near proximity of the terminus of my route,
if I had some one prepared to begin where I left off? It is not, however,
my good fortune to occupy that agreeable position. I shall continue the daily
record of my labors and the events that the times may bring forth. . . .
Gideon Lincecum

Lincecum arrived at Tuxpan[1] on June 16, 1868, with high hopes, and he
was not to be disappointed. His pioneering spirit responded to the
challenge of carving a homestead and farm out of a wilderness, in this
case a tropical one such as he had never before experienced. For the
next three years, his energies were so fully directed to the demands of
making a decent living for himself and his daughter's family through
his farm that he had little time to pursue his scientific interests. To his
daughter Sarah he wrote (June 11, 1871):

> When I look over my letters I am fully conscious that they contain
> but little that will interest you; yet I feel inclined to continue the record
> of my labors for the purpose of showing you the manner in which I
> employed the closing years of my life period. I, whose mind had rev-
> elled among the stars; played familiarly with the beauties and won-
> ders of natures secrets, and made them a matter of public record for
> years—starved and neglected the animal of my organism whilst in
> the eager pursuit of intellectual food—Am now fat and robust—starving
> the intellect, whilst I plant corn and potatoes and feed hogs to pet
> and feed and fatten the animal.

But in 1871 he was able to shift much of the burden of the farm
and even some of his medical work to Dr. George Bradford, who ar-

rived in Tuxpan and soon afterward married Gideon's granddaughter, Attilia Campbell.[2] Although Bradford soon tired of the backbreaking burden and retreated with his bride to Texas, Lincecum had a respite for a few months that enabled him to resume his passionate pursuit of nature's secrets with full gusto. The tropical birds and butterflies interested him immensely, as did the denizens of the Gulf waters and seashore. As always, his eyes were on the lookout for new species of his "brother emmets," the ants.

Many of the letters he wrote to family members from Tuxpan were like journals, and one might continue a month or more. In a letter started March 14, 1871, he described for Sarah Doran two chachalacas[3] that he shot on April 15, 1871: "They are about as large as a half-grown chicken with a long tail, color, dark brown; when picked they look very much like a picked chicken; . . . They are very numerous in our jungle. Very wild, requiring much caution and quick action in shooting them."

To his "Bully" grandson[4] in a letter dated July 13, 1871, Lincecum wrote about the butterflies migrating through the Tuxpan area:

Saturday 22 July, Large yellow butterflies[5] have been crossing the river from the north ever since the 16th. instant. For 10 or 12 days at this season of every year they may be seen passing in countless numbers. They come over and deposite their eggs on the leaves of the agua cate [avacado],[6] guava and other fruit trees, filling them with a very poisonous catterpiller or stinging worm.

Sunday 23—12 M. At this moment there are a greater number of butterflies coming over than any previous day. Grandpa will tell Bully that all the types of the butterfly kingdom are gorgeously painted and are very pretty things as beautiful and as varient as the flowers. Yet they are the most injurious, devourering, destructive thing and do more damage to our crops, and to the vegetable world generally than all other pests put together. The Locusta—grasshopper—is very bad occasionally, destroying wide districts in a single night, but for mischief he is not to be compared to the butterfly family, who are here always, producing all the voracious worms and catterpillers from the microscopic *Anacampsis cerealella*,[7] which secretes itself within a wheatgrain, devouring the mealy substance, upwards through a thousand types to the great blue winged beauty[8] of Tuxpan Valley, Mexico. It is while in the larvae state that they do the mischief. After they have taken on their painted habiliments and are afloat, they are perfectly harmless. The larvae of our great blue backed butterfly is five inches long, with warts bristling with red, green and black spines distributed at regular intervals along his back. They subsist on the tender twigs and leaves of trees. Fortunately they are not numerous.

Amongst the larvae of the butterflies a large number can sting severely, and a few of them are capable of destroying life. There is one here as bad as the rattlesnake bite.

To Sarah Doran (August 3, 1872) Lincecum reported the return of the yellow butterflies in 1872. "For ten days the large, bright yellow butterflies have been coming across the river in countless thousands. It is their time of the year for travelling." He sent several specimens in the letter and commented on them: "No. 5 is the above described butterfly. His larvae is large, yellow, with black bands, hairy, with very sharp spines, and possess an extremely virulent poison, which he emits from his spines on being touched. I have heard the children crying all night from the pain produce[d] from being stung by one of them. It is very bad."

In another letter to his daughter Sarah (April 29, 1871), Lincecum described the Portuguese Man-of-War:[9]

Saw many of the Portegee men of war come sailing ashore in great numbers. They are an interesting, and to me, quite curious affair. They are a jelly fish of perhaps, a pounds weight. Ovyform, having on their back a ruffled tumor, which they can inflate with air till it rises above its body 3 or 4 inches in the form of an arched cone; the lower part of which and the part of the body that is elevated above the water, is of a bright rich purple color. Having a fine effect on the vision of the

beholder as it comes sailing into port. The portion of the body that lies beneath the surface of the water has a ragged appearance, and projecting out from these fragment-like inequalities are long purple, strong strings of various lengths. Some of them two or three yards in length and as thick as a goosequill. It is said that when these rope-like appendages comes in contact with any ordinary fish, they instantly contract, wind around it, cord it up and produce so much pain by a poisonous fluid which is emitted from those strings that the fish not being able to extricate itself from the death dealing toils, soon expires, becoming food for the brainless sea pest.

When the wind is fair and they have their sail throughly inflated, it drives its ragged bottom and all those long strings very steadily on the water. If the wind is landwards, they are driven ashore, and they possess no power whereby they get to sea again. There they are, helpless on the beach; and, as the sun rises higher and hotter, they become more inflated until there is nothing left on the dry sand but a great blue globe, which, when trod upon forcibly, explodes with a loud pop. We know that the long strings that hangs in the water beneath this nondiscript creature can sting badly. Many of the sea bathers know it from painful experience.

In spite of such hazards, Lincecum enjoyed walking the beach and surf bathing, as he described in a letter to Sarah Doran, started April 29, 1871:

On Sunday 30 April—Set Sail at daylight in company with two other boats and families. . . . It is but 5 miles and we landed on the south side beach quite early. It is a pretty beach with its borders shaded with the banyan and uba (coba) trees.[10] . . . John [Campbell, his grandson] and I proceeded down the beach until we came to a stopping place; not exceeding three miles to where the sea and thorny jungle joined forces. Thorns and prickles on the right, and deep salt water on the left; it was plain that we were at the end of our journey. We commenced to retrace our steps, picking up pretty shells as we progressed, until we had filled our pockets; when we selected a suitable place, where we played in the briny breakers for at least an hour. It is frightful to look upon the foamy breakers—full of sharks and grampi[11]—as they come roaring and dashing their froth-capped head on the sea-beaten sands. Yet we not only looked on them, but we fearlessly plunged into, even the third wave; where being instantly overwhelmed with its mighty surges, [we] would soon be rolled up on the sandy shore again. To meet and encounter the third white topped wave in the best and most pleasurable style, you watch a receding wave and running down after it, dash through the first two you meet; you find yourself facing the relentless and irresistable third billow, which, without hestitation, receives you in its briny bosom, envelopes you in its frothy vestments, and whirling you heels upwards light as a feather, shorewards without regard to your posture or position, casts you on the sand.

This is certainly the most amusing, delightful & healthful recreation in the world. Octogenerian as I am, I enjoy it exceedingly. While swimming in and buffeting with the dashing billows of old ocean, I experience as I float the full glow of youthful vivacity, and I say to myself, "Am I not a boy again."

Lincecum also related the particulars about a fish kill in May, 1871, in a letter to Sarah Doran, begun April 29: "During the last 7 or 8 days, all sorts & sizes of fish have been coming ashore dead. The natives say that the sea grass berries are ripe; that the fish eats them, and that like the coculus indicus, it kills them.[12] All fish eating will be suspended now for 2 or 3 weeks; or till the fish gets well and fat again. . . . it is true about the middle of May every year, they die in astonishing numbers, so as to cause the entire sea beach to stinck with them. It is a deprevation."

During his stay in Mexico, Lincecum was not allowed to forget his former involvement in scientific investigation and correspondence. In a letter to Sarah dated April 2, 1871, he lamented:

Received by the Mexican post this morning, one letter from Beaumont, TX, and one from London, Eng. In addition to the postage stamps already placed on these letters, they are all taxed here and I have to pay twentyfive cents for every half oz. The Academies of Natural Sciences in all N. America and some in Europe are writing to and playing the devil with me. They put full postage on their letters. The imposition lies here. I shall have to leave them in the office, as the only remedy left me, to escape the illegal postage levied here. I have quit writing them, but they say they see my published letters and thereby know that I live.

Despite this protest, he continued to collect specimens of the animal and plant world, and he sent many to Sarah Doran as well as to his scientific correspondents. In a letter started on September 27, 1871, Lincecum wrote to Sarah that he had collected thirty-four species of ants and would send them to S. B. Buckley at a later time. He also sent Sarah a variety of seeds from the Tuxpan area, as detailed in a letter to her dated June 9, 1872:

In paper No. 1—is half a dozen seeds of the large white trumpet-shaped convolvulus.[13] I sent you a few in a former letter, and fearing they may

fail I send you these. Though the flowers are very numerous, they mature but few seeds. I am very desirious that you may succeed with these for they are the worlds wonder!

No. 2. Roman Cammomile.[14] . . . It is the anthemis Nobiles of the shops. Here it comes up about first of January, and the seed are now ripe & the plant is drying up. Sow some of the seeds in November and some others about the 10th of March. They are useful.

No. 3. Is the Yerba dulce.[15] . . . I hope you may get this sweet little arromatic herb to grow in your garden. It is as good and in many cases preferable to the Lemon balm or any of that class of medicinal plants.

In a letter to Sarah dated August 3, 1872, Lincecum enclosed additional seeds. These included:

Paper No. 1 is a vine of the bean family with large, very white pea blossoms. There are blue ones of the same kind here, no ripe seeds yet. I will watch them.

No. 2. is a small umbrageous tree,[16] bearing many nondiscript, white floweres. It is a pretty tree bearing beans. The spanish name of No. 1 is Zapatito de la reina,[17] which is being translated: The Queen's little shoe. . . . I think I sent a lot of the seeds of No. 2 some time ago. That however, will make no difference. Preserve and plant them after 20th. March. You will have to keep all the seeds I may send you during the remainder of this year, out of the ground till next spring.

Later in the letter, Gideon described two more plants:

. . . brought home two species of the guava, large red agua cata [avacado] and some watermelons. As I discribed to you in a former letter, the agua catta is beat to a soft pulp, salted and eat with other article[s] of

your meal, dipping it with your knife like butter; and it is most assuredly a better article of food. The seed of the agua cata is as large as a
hens egg. I could not send it to you, and though it is a large tree 6 or
8 feet in circumference sometime, I do not think it would survive a
Texas winter.

I shall send to you the seeds of both the guavas. No. 3 is the red

meated, pear shaped guava. No. 4 is the yellow pear shaped guava.
They are both very delicious fruit. It is said by the natives that in the
guava genus, there are seven distinct species. I have seen five. They
differ in color and shape. Very little difference in flavor. They are all
equally acceptable to me. . . .

No. 6. is a species of cactus. Its fruit is large and of a palish green
inside. Its flavor puts you in mind of the watermelon, but better than
the melon. I consider it a very delicious fruit indeed. I have seen nothing but the fruit; it is brought from the interior. No. 7 is the Chacloco[18]—
Mex. — It is a pretty shrub with capitate clusters of tubular, red flowers;
berries black; the whole plant has a purplish red appearance.

No. 8. Is the Bonnet gourd,[19] as some call it, Dishrag squash. I
have forgotten the botanic name of it. I have seen in time of the war
of 1812 in Georgia, very good looking bonnets that were made of it. To
scour furniture and for dishrags it is generly spoken of.

Lincecum also sent Sarah a dwarf china and said that "it will perhaps
require some protection during your cold northers."

In return, Lincecum's "Bully" grandson shipped seeds south to his
grandfather as evidenced in a letter started on July 13, 1871. Lincecum
stated on July 29:

I was proud of the Seeds you sent us, but was sorry to find them all
bored by the wevils; and the worst of the story is the wevils were,
quite a number of them, alive and kicking. I killed all I could get my
hands on, but some got away for which I am sorry; for we had none
of that kind of wevil here, and I am afraid it will introduce them to
our beans in this country. That would be a regrettable thing, for at
least one forth of the solid food of our people consists in a black
bean—frijoles, which is, assuredly the most substantial & universally
eaten article of diet we have.

Also, Sarah sent Gideon turnip and lettuce seed (August 3, 1872).

In a letter to "Bully" November 3, 1871, Lincecum gave an interesting
account of an invasion by what he called the "Bravo ant," apparently a
genus of *Eciton* or army ant:

Today about sunset the Bravo Ants appeared in the hog-pen and soon
after supper they came into my room. We were glad to see them and
to keep them off the beds, we immediately tucked down all the mosquito bars. We then kept out of their way and observed their action.
Their plan of invasion and success in capturing the insects that had
taken winter quarters in the house was not only ingenious but extensive and wonderful to witness.

Great numbers of them were dispatched from the main Army

into the house, where they were seen rapidly running over the walls in all directions and into the cracks and crevices and all the hiding places to be found, routing out every species of insects. It was amusing to see with what trepidation the insects fled from the approaching foe. While this was going on inside the house, the main body, in countless thousands, spread themselves on the ground outside like a carpet, in a belt three or four feet wide.

Soon the frightened roaches, ants, stinking pumpkin-bugs (which we could smell), spiders and many other creeping things were to be seen precipitately escaping from the invading foes that were scouring every part of the house. The ants inside did not capture many of the insects, for it seemed to be orders to drive them out to the devouring hosts, who were ready to sieze and devour everything that came out. By daylight they had all disappeared.

Lincecum had not expected the Texas northers to follow him to Tuxpan, but follow they did, and his correspondence made frequent references to their visitations, which often brought welcome cooling. At times the northers wreaked havoc on the small fleet of vessels that were Tuxpan's main connection with the outside world.[20] For example, in a letter begun March 14, 1871, he described for Sarah Doran a norther that came through on March 31: "The norther this morning has destroyed our corn extensively. We wonder how much harm it has done to you poor Texans. By tomorrow morning, you will have blasted fruit, frozen gardens, and ruined cornfields to contemplate."

Just a few weeks later, in a letter begun April 16, 1871, Gideon described another norther that reached Tuxpan on April 23:

We had a blustery norther two days. It run all the shipping out from the moorings. Vessels are afraid to remain at anchor during a strong norther in our moorings, and they run out into the open sea. It is several days sometimes before they get back to the anchorage.

For the most part Lincecum found the climate thoroughly to his liking and gloried in Tuxpan's natural setting. In a letter to Sarah Doran (March 14, 1871), he described Tuxpan in the spring of 1871: "How beautiful and pleasant are our spring mornings. The pretty river; the sweet health giving early breeze; the rain-like rattle and grand display of the bright green leaves of luxurient banana; and the substantial promise in the wavy cornfield, now in full bloom; the thick growing sugar cane, full of sweetness, with the ringing songs and love talk of the countless birds in the surrounding blooming forests; all combine to make life agreeable—to increase the bliss of being with all who possess the soul principle." In a letter to "Bully" (July 4, 1871), Lincecum responded to a request from his son Lysander[21] (San), who wanted Gideon to return to Texas and, as Gideon described it, "live with him, do nothing, have everything I want and travel 3 months every year away out on the Rio Grande. Tell him that I would do so with pleasure, was it not for the fact that I should dislike to exchange the beautiful, evergreen fruitful valley of the Tuxpan with its pleasant climate, delightful scenery and my good health, for the grassless, ruined prairies, dried up water courses, sun-scorched summers & terrible winter northers with its rheumatisms, croups and phthisics of Texas, to say nothing of the presence of the accursed Yankee & his thieving proclivities. Beg to be excused."

Although there were times when he felt keenly the absence of stimulating colleagues with whom he might discuss nature's secrets and revelations, he expressed no desire to return to Texas, despite missing the family members he had left behind. Only when it became clear that his favorite daughter, Sarah Doran, who shared his love for the natural world, could not visit him except at the expense of her growing family responsibilities, did he finally decide to pay a visit back to Texas. Around the first of June, 1873, he took a steamboat from Tuxpan to Galveston, intending to return to Tuxpan after making the necessary visits among family and friends. One of the latter was his old colleague S. B. Buckley, with whom he spent several days and nights at his retreat outside of Austin.[22]

A combination of unforeseen circumstances protracted his sojourn in Texas for over a year, by which time his aged body let him down and held him captive on Texas soil with a form of paralysis. During this time he corresponded again with a number of his scientific friends and undertook a long series of autobiographical articles for *The American Sportsman* newspaper.[23] In March, 1874, he received a letter from the prominent naturalist Elliott Coues of Washington, D.C., that must have brought him great satisfaction. Coues wrote:

> Although personally a stranger to you, I am not so to your numerous interesting writings on the habits of animals, and the freedom of intercourse which all good naturalists encourage among each other will I doubt not be extended to me by you on this occasion.
>
> To introduce the subject of this letter let me promise that I am preparing an extensive and elaborate work on the Mammals of North America, to be a complete history, technical description & biographical. I shall work up the living species, while Prof. Marsh[24] will do the fossil ones—and together we hope to make a great work.
>
> My chief deficiency is in the matter of accurate and minute information respecting the habits of the smaller animals—the rats, mice, gophers, squirrels, shrews, moles, &c. &c. Having seen all your publi-

cations on these subjects, in the *Naturalist* & elsewhere, I am convinced that there is no one in the country who has paid more attention to these things than yourself, or who has studied more successfully. And I am anxious to avail myself of your investigations.

Can you not, in a some what systematic manner, furnish me with a number of biographies—just such articles as you have so frequently communicated to the Smithsonian? Anything, however fragmentary it may be, would be welcome. I should only be too happy to incorporate all your observations in my work. I shall in any event collate and edit pretty much all you have published, but probably you would prefer to go over the points again, while it would be better on all accounts that the matter should come from you directly, rather than be transcribed from already published accounts.

I need not add how fully & gratefully all contributions to the work would be acknowledged, and passed entirely to your credit. I hope to have an early and favorable reply from you. There is no particular hurry for the present—any time within a year would be in season for the work.

I have lately completed and am about to publish an elaborate monograph of the *Muridae*,[25] and you need hardly be told how very valuable were the numerous specimens which I found in the Smithsonian from you.

There is a notation on the Coues letter stating that Lincecum answered it on May 4, 1874, but the reply is not found in the Lincecum papers. When Coues published his monograph on *Muridae*, the family of rats and mice, in Volume XI of F. V. Hayden, *Report of the United States Geological Survey* (1877), he cited Lincecum's writings on "The Gregarious Rat of Texas" and the "Texas Field Mouse," which had appeared in the *American Naturalist*. Coues also noted Lincecum's essay on "The Wood Rat," which appeared in *The American Sportsman*. In addition, Coues listed three specimens of *Sigmodon* (the Cotton Rat), which Lincecum had contributed to the Smithsonian. In another of Coues's monographs in the Hayden Survey volume, *Saccomyidae*, he cited two more of Lincecum's articles, both on "Perognastus fasceatus" (the Pouched Rat), which had appeared in the *American Naturalist* and *The American Sportsman*. Coues also paid respect in this article to "the venerable Dr. G. Lincecum, of Texas."

Unfortunately, Lincecum did not live to see Coues's publications acknowledging his work. The octogenarian's robust constitution had begun to fail him in the spring of 1874, and on the 28th of November nature took Lincecum to the end of its course from life to death, as he knew it inevitably and properly would. On August 22, 1858, he had consoled his friend, Dr. Andrew Weir[26] of Harmony Hill, Texas, on the loss of his mate, in words that fit his own death:

> At these things we must not repine. There is no preventive or remedy for this species of bereavement. Let intelligent man study his own nature and sources of his affections, let him observe his relative condition and the springs of his actions, and he will soon discover the causes of his calamities. He will find that disorganization or physical death is an unavoidable appendage of animal life. That the very construction of his nature insures the certainty of a subsequent derangement. And that the primary qualities of all sensitive beings gradually leads to disolution. No organic perfectibility has been discovered yet, on the plane of animal existence, which is capable of excluding the anticipation of decay through the progressive operations of physical causes upon the constitution; nor is there any "panacea, or elixir of life," though it were capable of extending the period, that possesses the power to bestow ultimate duribility upon beings organized like ourselves. In the action and reaction; the ceaseless mutations—composition, decomposition and recomposition—which we see and know is ever rolling onwards as the universal order of all material things, we are assured that death—organic disolution—is as natural, as necessary, and happens as often as life; and that none but the intellectual coward will flinch from his share of it.

Lincecum's writings make it clear that he believed that after death he would continue on his own infinite journey as an indestructible atom of the universe, sure of science's truth, beauty, and perfection. To his daughter Sarah he had written (undated but apparently in 1871–72):

> Free, or radiant electricity, is that which we hear rumbling in the skies. It is matter reduced to its ultimate atomic condition. All organized matter is susceptible of being reduced to its electric atoms. It is found everywhere. I will start a thought for you.
>
> —Duration, Space, Matter, Motion, heat—Electricity—*developement*, all combined, produces in my organism the idea of infinite action—energy. Call it God if you prefer that term. Electricity, as has been already said is matter reduced to it[s] ultimate atoms. Free or radient electricity occupies and fills boundless space.—Ye Gods what an idea. No limits! No bounds! No ends! Around—above—below—all is infinite—and yet, as if to vindicate its eternal essence in an unfathomable sea of space, the minutest attom of matter maintains and preserves its integrity!!

We can only add, "Amen."

Appendix 1

Index of Notable Correspondents

Spencer Fullerton Baird (1823–87), a vertebrate zoologist from Dickinson College, was arguably the most influential and far-sighted scientist in the United States during the thirty-seven years (beginning in 1850) he was associated with the Smithsonian, first as assistant secretary under Joseph Henry and later as secretary. He presided over Smithsonian natural history during an extraordinarily active period in the reconnaissance of the West through several major surveys. To him fell the task of reporting the scientific results of the surveys and ordering, preserving, and distributing the wealth of material that came to the Smithsonian. His encouragement of others like Gideon Lincecum to submit their collections, as well as his own research and writing, earned him the designation "patron saint" of directors of the Smithsonian in the May, 1996, issue of *Smithsonian* magazine, celebrating the sesquicentennial of its founding. David Starr Jordan and Jessie King Jordan, "Spencer Fullerton Baird," *Dictionary of American Biography*, Vol. 1, Pt. 1, 513–15; Sorensen, *Brethren of the Net*, p. 43; I. Michael Heyman, "Smithsonian Perspectives," *Smithsonian*, 27 (no. 2, May, 1996): 22.

Samuel Botsford Buckley (1809–83) was born and educated in the North but did most of his research in the South. He came to Texas in 1860 and later became Texas state geologist. He and Gideon Lincecum became active correspondents and friends, and Buckley accompanied Lincecum on a research expedition to West Texas in 1867. "Samuel Botsford Buckley," *New Handbook of Texas*, 1: 803–804. Although they remained friends, Lincecum referred to Buckley in 1871 as "a selfish investigator." Burkhalter, *Gideon Lincecum*, pp. 196–97. The editors believe that was a reference to the younger man's tendency to appropriate the findings of others into his publications. For example, Buckley received

This small building, recently moved from the former Lincecum place to another location in Washington County, served as Lincecum's office, from which he penned thousands of pages of letters. Photo by Keith Carter

credit for being the first to publish an account of the Texas agricultural ant, but that article drew heavily on the investigations over a number of years of Lincecum and his daughter.

Edward Drinker Cope (1840–97) of Philadelphia was a precocious scholar of natural science who later was editor of the *American Naturalist* from 1848 until his death. He had an early interest in reptiles but turned more to vertebrate paleontology in later years, a field in which he was a leading authority and scholar. George P. Merrill, "Edward Drinker Cope," *Dictionary of American Biography*, Vol. 2, Pt. 2, 420–21.

Elliott Coues (1842–99), army surgeon and naturalist, became a leading ornithologist and an authority on mammals, especially of the family Rodentia. At the time he wrote to Lincecum, he was naturalist and secretary of the United States Boundary Commission, 1873–76. Witmer Stone, "Elliott Coues," *Dictionary of American Biography*, Vol. 6, Pt. 2, 302–303.

Ezra Townsend Cresson (1838–1926) of Philadelphia was one of the founders of the American Entomological Society and an officer for sixty-five years, forty-two of those years editing its *Transactions*. His influence in the field of entomology was great, and Lincecum had a considerable correspondence with him. Leland O. Howard, "E. T. Cresson," *Dictionary of American Biography*, Vol. 2, Pt. 2, 540–41.

Charles Darwin (1809–82) considered himself an entomologist, among other specialties, and a considerable portion of his published work dealt with insects. Thus it is hardly surprising that when Gideon Lincecum read *On the Origin of Species* in 1860 he wrote Darwin about the Texas agricultural ants, seeing in their behavior supporting evidence for the theory of natural selection presented in the book. In the years immediately following the publication of *Origin*, Darwin paid increasing attention to insects, corresponding extensively with entomologists and encouraging them to pursue evolutionary themes in their own observations. This probably accounts for his taking the time to answer Lincecum's first letter and then presenting a digest of the two letters before the Linnean Society, which resulted in the publication of portions of them in the *Journal of the Linnean Society* in 1862. His subsequent book *The Descent of Man* drew heavily from entomology. Sorensen, *Brethren of the Net*, pp. 197–98.

Sarah Matilda (Sallie) Lincecum Doran (1833–1919) was the eleventh of Gideon's children. In 1865 she married William P. "Sioux" Doran. More than any other of Gideon's children, she shared his love of scientific investigation and valued his writings. She worked diligently after his death to get portions of them published, such as the articles that appeared in the *Publications of the Mississippi Historical Society* in 1904 and 1906. It was her sons, Clyde Bryan Doran and Frank Lincecum Doran of Hempstead, who deposited the Lincecum Papers in the Center for American History at the University of Texas in Austin in 1930 and 1931. She and her husband (see below) are both buried in a Hempstead, Texas, cemetery.

William P. "Sioux" Doran (1836–1901) was born in New York State but migrated to Texas and supported the Confederacy in the Civil War, serving both as a soldier and as a war correspondent. He married Gideon

Lincecum's youngest daughter, Sallie, in December, 1865. Burkhalter, *Gideon Lincecum*, pp. 84–85.

Élie (but known as Elias) Magloire Durand (1794–1873) was born in France. Commissioned a pharmacist in the French army in 1813, he was a veteran of the Napoleonic Wars. He left France in 1816 because of his Napoleonic sympathies and in 1825 opened in Philadelphia a pharmacy which became an informal clubhouse for scientists. He was elected in 1824 to the Philadelphia Academy and to the College of Pharmacy and in 1854 to the American Philosophical Society. In 1868 he gave his collection of over 100,000 botanical specimens (including many obtained from Lincecum) to the *Jardin des Plantes*, Paris. George Harvey Genzman, "Élie Magloire Durand," *Dictionary of American Biography*, Vol. 3, Pt. 1, 538–40.

Cassandra Lincecum Durham (1832–77) was Gideon's fourth daughter, born in Columbus, Mississippi. She married George J. Durham in Washington County in 1852, and because he held several prominent political jobs in Austin, she was for a time a leader in that city's social elite. After the death of her husband in 1869 (while Gideon was in Tuxpan), she supported herself and her three children by operating a boarding house in Austin. Burkhalter, *Gideon Lincecum*, pp. 83–84, 294.

George J. Durham (1820–69) married Gideon Lincecum's daughter Cassandra in 1852. Born in England, he immigrated to the United States in 1835 and to Texas in 1837, settling in Austin, where he served in the local and state governments. An active naturalist, he became an authority in ornithology and viticulture. Burkhalter, *Gideon Lincecum*, pp. 82–83.

George Englemann (1809–84) was born and educated in Germany, came to the United States in 1832, and settled near St. Louis, where he developed a very successful medical practice. He carried on scientific studies in botany, meteorology, geology, and zoology that gained him a reputation as the leading scientific personality in the American West. He was an authority on western plants and a key correspondent of Asa Gray. Sorensen, *Brethren of the Net*, pp. 236–37.

Vice Admiral Robert Fitzroy (1805–65) was a distinguished officer of the Royal Navy, a member of Parliament, and a fellow of the Royal Society. He commanded HMS *Beagle* during Darwin's historic voyage, 1831–36. A hydrographer and a meteorologist, in 1854 he became chief of the meteorological department of the Board of Trade. John Knox Laughton, "Robert Fitzroy," *Dictionary of National Biography*, 7:207–209.

Col. Caleb Goldsmith Forshey (1812–81) took an active interest in education, science, military affairs, and railroad development. He founded the Texas Military Institute and headed it until the Civil War. He was a weather observer for the National Observatory. Burkhalter,

Gideon Lincecum, p. 180; James Roger Fleming, *Meteorology in America, 1800–1870*, pp. 27–30; "Caleb B. Forshey," *New Handbook of Texas*, 2:1084.

R. B. Hannay was an English visitor who was stranded in Texas by the Civil War. It is unclear how Lincecum came to correspond with him. Burkhalter, *Gideon Lincecum*, p. 160n.

Joseph Henry (1797–1878), despite having no college education, became a leading physicist and a professor of science at Princeton before his selection as first secretary of the Smithsonian Institution, where he had a profound influence on U.S. science, including meteorology. William F. Magie, "Joseph Henry," *Dictionary of American Biography*, Vol. 4, Pt. 2, 551–53; Fleming, *Meteorology in America*, pp. 20–24, 32, 64, 75–78.

Asa Gray (1810–88) was born and raised in New York State but for over forty-five years was professor of botany at Harvard. He was considered the leading U.S. botanist of his day and "one of the greatest botanists of the world." George H. Genzmer, "Asa Gray," *Dictionary of American Biography*, Vol. 4, Pt. 1, 511–14.

Joseph Leidy (1823–91) was considered "the foremost American anatomist of his time," but his interests and activity in natural science were wide ranging and his output was prodigious. He was a longtime and active member of the Philadelphia Academy. George P. Merrill, "Joseph Leidy," *Dictionary of American Biography*, Vol. 6, Pt. 1, 150–52.

John Peachland Lincecum (1820–1907), often addressed as "Nephew John," was the son of Gideon's brother Grant and lived with his father in Catahoula Parish, Louisiana. Gideon corresponded with him frequently as a means of keeping in touch with the family of Grant, to whom he refused to write because Grant was an ardent "religionist." Burkhalter, *Gideon Lincecum*, p. 114.

Mary Catherine Lincecum Matson (1825–?), the sixth of Gideon's children, was born in Mississippi. She married James V. Matson of Washington County in 1848, soon after Gideon's family arrived in Texas. Gideon had a great falling out with James Matson in 1860 and virtually disinherited Mary. Burkhalter, *Gideon Lincecum*, pp. 79–80.

Capt. Matthew Fontaine Maury (1806–73), a career officer in the U.S. Navy 1825–1861 (when he joined the Confederate Navy), was an avid scholar of navigation and meteorology and is today considered the father of oceanography. For many years he supervised the Naval Observatory and promoted gathering of hydrographic and meteorological reports on land and sea. Although Lincecum addressed him as "Lieutenant," his rank in 1859 was that of Commander. A controversial figure, Maury was disliked by some. H. A. Marmer, "Matthew Fontaine Maury," *Dictionary of American Biography*, Vol. 6, Pt. 2, 428–31; F. L. Williams, *Matthew Fontaine Maury, Scientist of the Sea*.

Henry Christopher McCook (1837–1911) was an Ohio-born Presbyterian minister and Union veteran of the Civil War. He was "an ardent naturalist," with a special interest in ants. In addition to publishing the *Natural History of the Agricultural Ant of Texas* (1879), he also wrote *Ant Communities and How They Are Governed, A Study in Natural Civics* (1909). Leland O. Howard, "Henry Christopher McCook," *Dictionary of American Biography*, Vol. 6, Pt. 1, 603; Burkhalter, *Gideon Lincecum*, pp. 222–24.

Daniel Boone Moore (1805–89) and his wife, Emily Lincecum (1813–84), who was Gideon's sister, received frequent letters from Gideon. They had married in Mississippi, where Emily had moved with her father, Hezekiah, and several of her brothers. During and after the Civil War they were living near Castroville, in Medina County, after pioneering earlier at Uvalde. Ibid., pp. 118, 154.

Alpheus S. Packard (1839–1905) was a well-educated native of Brunswick, Maine, and a Union veteran of the Civil War. At the time Lincecum wrote to him, he was the curator of the Essex Institute and the Peabody Academy of Science in Salem, Massachusetts. A founder and editor-in-chief of the *American Naturalist*, his specialty was entomology but, like Lincecum, his interests were broad. Leland O. Howard, "Alpheus Spring Packard," *Dictionary of American Biography*, Vol. 7, Pt. 2, 126–27. Lincecum's letters to Packard can be found in the Director's File of Letters Received, 1866–68, at the Peabody Essex Institute in Salem.

George W. Peck (1837–1909) was a New York City businessman as well as "a serious student of entomology," and a longtime member of the American Entomological Society. Butterflies and moths were his specialty. Sorensen, *Brethren of the Net*, p. 183. In response to his request for Texas specimens, Lincecum's diary for August 28, 1867, indicates that on that day he finished arranging a collection of butterflies to send to "Mr. Peck of N. Y. There is 1203 specimens, all in fine order and well conditioned."

Capt. Ben Shropshire was a respected lawyer, newspaperman, and district judge of LaGrange, Fayette County, Texas. Julia Lee Sinks, "Editors and Newspapers of Fayette County," *The Quarterly of the Texas State Historical Association*, 1 (July, 1897): 37.

Benjamin Franklin Shumard (1820–69) was another Northern scientist who came to Texas before the Civil War but returned to the North when the war began. Trained in medicine, he developed a strong interest in paleontology and geology and served as Texas state geologist, 1858–60. "Benjamin Franklin Shumard, *Handbook of Texas*, 2:608; *New Handbook of Texas*, 5:1037–38. A rival of Lincecum's friend S. B. Buckley, Shumard incurred the contempt, probably undeservedly, of Lincecum. Nevertheless, the two men were on good terms prior to the controversy over Buckley, and Shumard named a Texas fossil after Lincecum. Burkhalter, *Gideon Lincecum*, pp. 184–85.

James Webb Throckmorton (1825–94) gave up medicine for politics and served Texas as a military officer, governor, and congressman. Although a moderate and former Unionist, he was removed as governor by the Federal Reconstruction authorities on July 30, 1867. "James Webb Throckmorton," *Handbook of Texas*, 2:778; *New Handbook of Texas*, 6:485–86.

Horatio Charles Wood (1841–1920) was a productive scholar and teacher in medicine and natural history. Born, raised, and educated in Philadelphia, he taught botany, pharmacology, and medicine at the University of Pennsylvania. He also had a strong interest in entomology and accompanied several Smithsonian expeditions beyond the borders of the United States. Joseph McFarland, "Horatio Charles Wood," *Dictionary of American Biography*, Vol. 10, Pt. 2, 459–60. Lincecum underestimated Wood's talent and misjudged his character. Although Lincecum addressed him as H. C. Wood, Jr., the father's name was Curtis, not Charles.

Appendix 2

The Edited Version of Gideon's Letters to Charles Darwin That Appeared in Journal of the Linnean Society of London (Zoology), 6 (1862): 29–31

Notice on the Habits of the "Agricultural Ant" of Texas ["Stinging Ant" or "Mound-making Ant," *Myrmica (Atta) malefaciens*., Buckley]. By GIDEON LINCECUM, Esq., M.D. Communicated by CHARLES DARWIN, Esq., F.R.S., F.L.S. [Read April 18, 1861.]

The following is merely an abstract of Dr. Lincecum's communication, containing only what appears to be most remarkable and novel in it in the way of observation.

"The species which I have named 'Agricultural,' is a large brownish Ant. It dwells in what may be termed paved cities, and, like a thrifty, diligent, provident farmer, makes suitable and timely arrangements for the changing seasons. It is, in short, endowed with skill, ingenuity, and untiring patience sufficient to enable it successfully to contend with the varying exigencies which it may have to encounter in the life-conflict.

"When it has selected a situation for its habitation, if on ordinary dry ground, it bores a hole, around which it raises the surface three and sometimes six inches, forming a low circular mound having a very gentle inclination from the centre to the outer border, which on an average is three or four feet from the entrance. But if the location is chosen on low, flat, wet land liable to inundation, though the ground may be perfectly dry at the time the ant sets to work, it nevertheless elevates the mound, in the form of a pretty sharp cone, to the height of fifteen to twenty inches or more, and makes the entrance near the summit. Around the mound in either case the ant clears the ground of all obstructions, levels and smooths the surface to the distance of three or four feet from the gate of the city, giving the space the appearance of a handsome pavement, as it really is. Within this paved area not a blade of any green thing is allowed to grow, except a single species of grain-bearing grass. Having planted this crop in a circle around, and two or three feet from, the centre of the mound, the insect tends and cultivates it with constant care, cutting away all other grasses and weeds that may spring up amongst it and all around outside of the farm-circle to the extent of one or two feet more. The cultivated grass grows luxuriantly, and produces a heavy crop of small, white, flinty seeds, which under the microscope very closely resemble ordinary rice. When ripe, it is carefully harvested, and carried by the workers, chaff and all, into the granary cells, where it is divested of the chaff and packed away. The chaff is taken out and thrown beyond the limits of the paved area.

"During protracted wet weather, it sometimes happens that the provision stores become damp, and are liable to sprout and spoil. In this case, on the first fine day the ants bring out the damp and damaged grain, and expose it to the sun till it is dry, when they carry it back and pack away all the sound seeds, leaving those that had sprouted to waste.

"In a peach-orchard not far from my house is a considerable elevation, on which is an extensive bed of rock. In the sand-beds overlying portions of this rock are fine cities of the Agricultural Ants, evidently very ancient. My observations on their manners and customs have been limited to the last twelve years, during which time the enclosure surrounding the orchard has prevented the approach of cattle to the ant-farms. The cities which are outside of the enclosure as well as those protected in it are, at the proper season, invariably planted with the ant-rice. The crop may accordingly always be seen springing up within the circle about the 1st of November every year. Of late years however, since the number of farms and cattle has greatly increased, and the latter are eating off the grass much closer than formerly, thus preventing the ripening of the seeds, I notice that the 'Agricultural Ant' is placing its cities along the turn-rows in the fields, walks in gardens, inside about the gates, &c., where they can cultivate their farms without molestation from the cattle.

"There can be no doubt of the fact, that the particular species of grain-bearing grass mentioned above is intentionally planted. In farmer-like manner the ground upon which it stands is carefully divested of all other grasses and weeds during the time it is growing. When it is ripe the grain is taken care of, the dry stubble cut away and carried off, the paved area being left unencumbered until the ensuing autumn, when the same 'ant-rice' reappears within the same circle, and receives the same agricultural attention as was bestowed upon the previous crop,—and so on year after year, as I *know* to be the case, in all situations where the ants' settlements are protected from graminivorous animals."

In a second letter, Dr. Lincecum in reply to an inquiry from Mr. Darwin, whether he supposed that the ants plant seeds for the ensuing crop, says, "I have not the slightest doubt of it. And my conclusions have not been arrived at from hasty or careless observation, nor from seeing the ants do something that looked a little like it, and then guessing at the results. I have at all seasons watched the same ant-cities during the last twelve years, and I know that what I stated in my former letter is true. I visited the same cities yesterday, and found the crop of ant-rice growing finely, and exhibiting also the signs of high cultivation, and not a blade of any other kind of grass or weed was to be seen within twelve inches of the circular row of ant-rice."

In his second letter Dr. Lincecum proceeds to give some account of what he terms the "Horticultural Ant," which appears to be identical with the "Cutting Ant," *Œcodoma mexicana*, Sm., described by Mr. S. B. Buckley in the 'Proceedings of the Academy of Natural Sciences of Philadelphia,' 1860, p. 233; but as his account does not contain any important additional observations, it is here omitted. Mr. Buckley also describes the "Agricultural" or "Mount[d]-making Ant," although his account of its habits will be found to differ in several respects from that given by Dr. Lincecum.

Appendix 3

Thirteen Distinct Species of Texas Grasses by Gideon Lincecum from Texas Almanac (1861), pp. 140–43

NO. 1. ONE OF THE WILD OATS. PERENNIAL.

This is seldom found in the prairie, which shows that it is a favorite with the cattle and is all eat up. But in the locks of the fences, where the cattle can not reach it, it is found growing large, three to four feet high, notwithstanding that it is in such situations surrounded with thick weeds and grasses of other kinds. It shows itself to be a fine meadow-grass, and capable of producing abundant crops of most excellent and very sweet hay, and as it matures by the 15th of May, comes in good time for the work-horses. I have seen thirty full-grown heavy-ended stocks to one root. I consider it one of our best grasses for hay.

NO. 2. BIENNIAL.

This grass in ordinary situations, where it has sowed itself, along the roadsides and places where the other grasses and weeds have been eaten out, on account of its being too thick, does not grow exceeding six inches high; but in low, moist lands, and when the ground is not over-cropped, it rises two or three feet in hight, and makes good hay. It comes up in November, is green all winter; is grazed on by hogs, cows, horses, sheep, and is ready for the scythe by the middle of May. After it has been mowed down, the stubble dies like wheat-stubble. I have seen twenty stems on one root.

NO. 3. BIENNIAL.

This is the smallest of the Rye genus. Like No. 2., it is found taking possession of the eat-out places, road-sides and locks of the fences. Like it, too, from being too thick, is seldom found more then six inches high. This and No. 2 are nearly always found together; in their habits, size, choice of locality, and the odor of the hay that is made of them, so much alike, that they may be estimated at about the same value. Two to three feet; matures in May.

NO. 4. WHEAT GRASS.

This grass comes up from the seed in November. In January, February, and during the spring it has the smell, taste and general appearance of wheat; horses, cows, etc., graze on it as they would on wheat. When it heads up it is about as high and has the same appearance, but its grain is precisely like flax seed, and falls out very early when ripe. I cultivated two acres of it two or three years ago, and cut it down about the last of April. It produced a fair quantity of the best hay—it was, in its nature, more like good fodder, and the horses ate it freer than any I ever had. I think it superior, when properly put in the ground, to rye or barley, for winter pasture.

NO. 5. RESCUE GRASS.

This grass is found in all kinds of soil west of the Brazos, is a biennial, indigenous plant, and will yield heavy crops of hay when rightly managed, but it is inferior to several other species of our native grasses. Recently it has been much talked of in Georgia and Alabama, and other Southern States, and not without some pretty good reasons; but I think, when compared with a good many species of our Texas grass, it has been overrated. It, however, is a very good grass, three to four feet high, and matures about the middle of May.

NO. 6. BIG MESQUIT.

My meadow, which is now ten years old—really, it is as old as the prairie, for the ground has never been plowed, but has been inclosed for ten years—had but very little of this species of grass at first. It is about half of that kind now. Its roots are triennial, and it produces good nutritious hay, in great quantities. My horses, mules, oxen and milk-cows are fed on it every winter, and they do exceedingly well upon it. Higher up the country vast tracts of good prairie lands are found heavily coated with this grass alone, producing excellent summer range for all kinds of stock.

NO. 7. WINTER GRASS. PERENNIAL.

This is superior to any grass I have yet seen in any country. For winter pasture it has no equal. It will flourish finely in any of our ordinary post-oak lands, is very green all winter, and is devoured voraciously by all the graminiverous animals; hogs eat it freely.

When cultivated for a meadow, it should not be grazed off during the winter, as its long, juicy, winter leaves make the very best kind of hay, when mowed and properly cured in the spring. It is headed up and ready for the scythe by the last of April; two and a half to three feet high, and is a very superior grass for sweet, nutritious forage; just smell of it now, while you have it in your hand. I am not certain, but I think its roots are triennial.

I think it belongs to the Agrostis family, and I have ventured to name it, *A. Texaria*. I have not, however, studied the botanical character of the grasses with sufficient care, to be satisfied that I can correctly place its generic name, in a strictly arranged, scientific nomenclature. This year, I have put up carefully prepared specimens of it, as well as several other kinds of our fine native grasses, which will be sent to the Academy of Natural Sciences, Philadelphia, where they will be analyzed and receive their permanent characters.

It may not be a subject of much importance to the farming portions of the community, but it does seem to me that a little attention to the great variety of our indigenous forage grasses would remunerate the effort satisfactorily.

NO. 8. GAMMER GRASS. INDIGENOUS.

This is a large, strong-growing grass in Texas, delights in moist situations, but will produce two heavy crops of hay in one season on any of our black prairie lands. Nearly one half the grass in my meadow is the Gammer. It produces rather rough, but if mowed early, very excellent winter forage for mules, oxen, and milk cows; horses eat it freely, and do pretty well on it. A great deal has been said about this grass several years past; it should be pretty well known, and I will refrain from further description.

NO. 9. BARBED MESQUIT. PERENNIAL.

This is the species of grass that attracts the attention of the traveler, and that we hear so often spoken of in Texas. Twenty years ago very little of it was seen east of the Colorado, but it is now found as far east as the Trinity River. It is very rapidly progressing, eastwardly, at least. It is a very excellent winter grass, very similar in appearance to the Blue grass. It is, during the winter, much sought after by stock of all kinds. Swine, where it is plenty, keep in good order by grazing on it. When the spring sap rises in it, cattle refuse to eat it; hence the cause of its spreading so rapidly. It is not prevented from maturing a full crop of seed, which it does, and casts them down by the last of May. It is a fine meadow grass, two or three feet. April.

NO. 10. HOG-WALLOW MESQUIT. PERENNIAL.

Before Texas was settled up, and the prairies considerably eat out, this species of grass was found only in hog-wallows; hence its name. It is now found, not only in hog-wallows, but is rapidly spreading itself along the roadsides and carpeting all the old roads and other spots and places of ground which have been denuded of other grasses with a thickly crowded coat of extremely fine, nutritious pasturage for summer grazing, for every type of graminivorous animals. In appearance, as it lies spread out on the ground, it very closely resembles the Bermuda grass, *Cynodon Dactylon*; like it, also, in its having two modes of propagation, by producing seeds and by creeping, taking root at the cane-like joints of its prostrate off-shoots. Its inflorescence and fructifying processes are widely different, as it does not belong to the same genus. Except on suitably moist grounds, where the stock can be kept from it, it is not large enough to make hay of, as it is not, on ordinary soil, exceeding three to six inches high, yet it affords good and frequent grazing. For this purpose it has no equal, and I will here venture the prediction, that the time will come, in Texas, when it will be thought more of, small and insignificant as it may now appear to the superficial observer, than any other of our indigenous grasses, for the purposes of summer pasturage. It is more especially adapted to the habits and peculiarities of the sheep, than to any other animal, and it seems to enjoy it more; yet all the grass-eating races devour it without hesitation and with good gusto.

NO. 11. (HAS NOT BEEN NAMED.)
PHLEUM? [sic] PERENNIAL.

Grows best in moist lands, and is, I think, a valuable meadow-grass. The root from which the specimen was taken, had forty-two well headed stems, thirty inches high. I notice that the cattle eat it entirely up outside of the inclosures, and that is one of my tests for ascertaining the best grasses. It produces abundant crops of seed, and can be easily propagated to any desirable amount.

NO. 12. CROW FOOT. BIENNIAL.

This is a thrifty growing, large grass in moist situations; will make very excellent pasture grass, and is easily propagated; but, on account of its thick, juicy leaves, would be difficult to cure properly. It has strong roots, spreads finely, and from the greedy manner in which all the graminiverous animals devour all they can find of it, its reputation as a good pasture grass is already established.

NO. 13. PERENNIAL WILD RYE.

Of this there are two species, one of them biennial, found in bottom lands. They are equally valuable for hay. The specimen (perennial) should perhaps have the preference, as it flourishes well on any of our ordinary uplands, and would not require seeding the ground more than once in three years.

Appendix 4

Articles on Geology by Gideon Lincecum from Texas Almanac *(1868), pp. 85–91*

1. GYPSUM IN TEXAS

This is the sulphate of lime, or plaster of Paris. It contains twenty-one percent of water. It is found crystallized in broad foliated plates, and also in compact masses. It is ground while in a natural state, and spread on certain kinds of land as a manure. The compact kinds are called alabaster, which is made into various ornamental articles. Some specimens are perfectly white, and, being translucent, are very beautiful. By being colored with the metallic oxides, other varieties present a clouded, striped, or spotted appearance.

Sulphate of lime is extensively employed in forming the ornamental or stucco work for public buildings. The most approved method of preparing it for this purpose is, first, to heat it nearly to redness, which is done to expel the water of crystallization, when it is ground in a mill. It is now a fine white powder, which, being mixed with water and cast into moulds of various figures, forms the ornamental works seen on the walls of public buildings.

Sulphate of lime is soluble in about five hundred parts of cold water, and it is more frequently found in the water of wells and springs than perhaps any other salt. When in considerable quantities, the water is called hard, and will decompose soap, and therefore is unfit for washing.

There are a great many varieties of it found in Texas. Some specimens are as transparent as the purest glass, and easily split into thin layers, like isinglass, or mica; others, again, are found varying in color;

being stained with metallic oxides, it represents many shades, from a pure transparency down to a dull, earthy color. Many portions of Texas abound with it in its compact form; and when our people shall become familiar with its various uses and applications in the arts and husbandry, it will be extensively sought after and used in buildings, castings, bust, ornamental vases, and fancy figures of every description.

This valuable mineral is found in greater or less quantities throughout the cretaceous formation of the State. In quarrying the living sandstone, it is often seen in small, odd-shapen pieces, resembling flint, but easily shaved with a knife; generally more or less transparent, and frequently stained with a yellow hue. In these situations it seldom occurs in sufficient amounts to remunerate the laborer, and it is referred to in this place only for the purpose of instructing the uninitiated.

In many sections of the State it occurs in large deposits. Two miles above Round Rock, on Brushy creek, Williamson county, it crops out in the form of large rocks, and is pretty pure. At the falls of the Brazos, in Falls county, considerable quantities, of a rather dirty-looking variety, but excellent for manuring lands, is seen projecting from the old volcanic cracks in an immense bed of blue shale, which underlies a large district about the falls and westwardly. It is over a break in this rock that the Brazos river falls, and it is fast wearing away under the action of the rushing waters. All the sand-bars for some distance below are literally covered with scrap[s] and chunks of gypsum, of various sizes, which have been dislodged by the caving and tumbling down of the ever-receding rock. The variety of this article that is deposited at this place would, when properly prepared, produce excellent plaster for building purposes and for fertilizing poor lands. That near Round Rock is of a more compact and finer quality, and, when prepared by a proper degree of heating and grinding, will produce a fine plaster for casts, fine mouldings, and other ornamental works.

All around Mount Bonnell, near Austin, on its top, and in many of the beds of lime-rock in the vicinity, there are quantities sufficient to supply the demand of all the builders and plasterers of that city. On Cowhouse creek, in Coryell county, there are large amounts of it; in short, it is a conspicuous ingredient in the rock composition of many portions of the State, and in some places abounds in large bodies. Higher up the Texian rivers, in the region where they cut through the mountains and penetrate the abruptly elevated edges of the great Llano Estacado, it occurs in such abundance that it impregnates all the waters of that region with its bitter principle, rendering it entirely unfit for use. Horses and other animals refuse to drink it.

Captain Marcy, in his exploration of the Red river of Louisiana, states that the peculiar bitter taste of its water is communicated by ingredients that it receives in flowing for a hundred miles over a gypsum bed; and that this gypsum range forms a belt of immense extent, which crosses the country for some four or five hundred miles. "It is," he continues, "regarded by Dr. Hitchcock as the most extensive deposit of this mineral in North America." It is everywhere characterized by the same peculiarities, with the water issuing from it invariably bitter and unpalatable. The Arkansas, Canadian, Brazos, Colorado, and Pecos rivers pass through this formation, and a similar taste imparted to the waters of all. The greater portion of this vast belt of sulphate of lime—plaster of Paris—which is at least 100 miles wide, lies within the limits of Texas.

The greater quantities of plaster of Paris that are annually made away with for one purpose and another in the United States show how very highly it is estimated, and that it is a very popular article in commerce and social economy; and it has been satisfactorily demonstrated that within the limits of our State there is plenty of this almost indispensable mineral—some of it in nearly every county—to supply the demands of our people for building purposes thousands of years, and to cheer the spirits of the coming generations with the joyous certainty that they possess the means, at a cheap rate, of reclaiming the vigor of their waning lands with a capital fertilizer, if their future exhaustion should require it.

It is a subject of wonder to me why the managers of our public buildings should prefer to send off to Europe and other countries for plaster, when there is such an unlimited supply of as good, if not a better article, near at hand in our own country, which can be obtained for half the cost of the foreign article.

In the future, some intelligent mind, with small capital and big energy, will go to work with this highly useful, cleanly, fashionable, and superbly ornamental mineral, and soon the indications of his calling will be conspicuously manifested in every tasteful dwelling in the land.

2. TEXAS MARBLE

The marble falls on Colorado river, fifty miles above Austin, is about the centre of the marble region, and is the most attractive and interesting locality in the State. It is indeed a most delightful place, affording beautiful waterpower in sufficient force to drive immense amounts of machinery. The upper falls are crescent form, three in number; and the rock over which the water tumbles is quite black, capable of bearing a fine polish, and is called marble by the people of that vicinity. In the course of a mile and a half, there are many little falls, over which the water descends, in the aggregate about sixty feet. The uppermost falls are all black rocks. The middle falls consist of a series of ledges of very pretty gray marble, and at the lower falls the water pitches over thick strata and large blocks of very fine white marble, which in quantity is inexhaustible. The dip of these rocks is about 20° west. The valley above the falls is twelve to fifteen miles in length and five or six miles wide. It is the bed of an ancient lake which is bounded on three sides by a circular mountain of about two hundred feet elevation. At the swag of this half-circle, at the point where the base of the mountain is four or five miles across, the waters of the lake, inflowing over its top, made a breach which in the countless centuries of the past have whetted away the mountain and its marble entrails until the *now river* has imbedded itself down into the marble base at least two hundred feet. In other words, this extensive obstruction, this mountain dam, has been torn in two, and its solid marble base ripped up by the dashing waters and rolling pebbles of the comparatively little river.

As proof of the hypothesis that the valley immediately above the falls is the bed of a former lake, it is only necessary to state that everywhere in it where the plow has been employed, shells in a perfect state of preservation, of the same species of the Unios that inhabit the river now, have been turned up in considerable numbers. The soil of the valley is mostly sandy. Generally it is fertile, producing good corn, wheat, and the ordinary crops of the country.

To return to the falls, the upper series of which, as I have already stated, are in the form of a crescent, composed of three shelves or layers of unequal extent, of the black rock, dipping toward the upper river, west about 20°, altogether constituting a fall of twenty-five or thirty feet. Up the inclined plane formed by the dip of these rocks, the pebbles, small boulders, and the *debris* of the mountain above have been driven by the swift-running river for a period of many ages, until deep channels have been worn in the rock through which the rushing waters now pass, forming perhaps a hundred beautiful cascades, over which a footman can, by stepping these narrow, foamy, water-passages, very conveniently cross the river dry-footed. These numerous and most beautifully arranged little dashing water-sluices are confined to the upper stratum of the black rock series. In the two lower strata similar channels have been cut, but, in consequence of a gap of fifteen or twenty feet torn out near the centre, the water of the entire river gathers to this point and rushes through with great force.

I have already remarked that the marble falls are situated near the centre of the marble region. In Hamilton's Creek Valley, three miles below the town of Burnet, is a quarry of fine marble, which, having a crooked, purple, thread-like marking blended through its composition, gives it a peculiar and rather handsome appearance. It was from this quarry that the marble block sent by Texas to the Washington Monument was taken. It was spoken of at Washington as being a specimen of superior quality. From the Hamilton's Valley deposit, over the extremely rocky hills and little mountains, lying in a southerly direction, to the distance of thirty-five to forty miles, including the marble falls on Colorado, are immense deposits of hard limestone and a great deal of fine marble. In this stony region is occasionally seen a small cabin constructed of little crooked poles, the longest they could get not exceeding fourteen feet, with an extravagant marble chimney to it. Their little farms are sometimes fenced with marble. When, at the marble falls, its great and easily available water-power shall be brought into requisition, it will be found amply sufficient to drive all kinds of machinery at the same time. Besides the cotton and woolen mills, grist-mills, and saw-mills, there will be machinery to cut, dress, and polish the fine marble, and there will be many other things manufactured there that have not been thought of yet.

For its fine white lime that region will be looked upon as the Thomaston of the South, and its exceedingly fine classes of marble will attract the attention of the civilized portion of the earth. It will be sent to every clime, and the rattle and the clatter of the great works that shall be set in motion there, the hum of business and the manifestations of the industrial activity that shall be seen there, will astonish the visitors and traders from every country.

The objectors to progress need not allege the lack of facilities for transportation. Capital can and will reach all such rich deposits and natural stores of wealth, with its railroads; and I predict now, on this 4th day of July 1867, that many of those little minds, whose mechanical sagacity is incapable of penetrating the rock-ribbed mountains where the tunneled railroads are to go, will live to see and wonder at the stupendous works soon to be put in operation.

3. THE MEDICATED WATERS OF TEXAS

The mineral springs of Texas are very numerous, and they are pretty equally distributed over the State. To describe the localities, chemical properties, and the healing powers of all of them is what I shall not attempt to accomplish in this paper. I am, however, familiar with a few of them, and to them I shall confine my remarks.

The first and perhaps the greatest of these health-giving waters is the Lampasas Sulphur Springs.[1] These springs gush forth from the split rocks in the bed of the sulphur fork of Lampasas river, from its bank and in its valley to the extent of eight miles along its course. They are called white sulphur springs, and from the eagerness manifested by the cattle to obtain the water from them, they doubtless contain a considerable percentage of the chlorine of soda (common salt), which may add greatly to its powers as a disinfectant and depur[g]latory agent.

The uppermost one of these springs rises from a split rock in the bed of the creek; it is a kind of basin, and it is filled up to a level with the surrounding rock floor with small boulders or pebbles, composed of every species of stone common to that vicinity, and these are being rapidly worn and polished into all manner of shapes and forms, from being incessantly thrown to the top by the gushing waters. This spring throws out, according to the best estimate I could make, two barrels of water every second. Many bushels of pebbles are perpetually dancing and audibly tinkling in the upward rush of this great column of water. Nothing can sink in it. A child or stone, or anything else thrown into it is instantly hurled out again. I have seen hairpins and finger-rings that had been dropped by the ladies while bathing picked up amongst the whirling pebbles. Fifty or sixty yards from this wonderful fountain of water, situated in a basin forty feet in diameter, in the valley of the creek is another very flush spring of the same kind of water. It does not discharge exceeding one twelfth the quantity of water that flows from the big spring. The smaller spring is the one from whence the invalids dip the water they drink, using the larger one to bathe in. On both sides of the creek, in the vicinity of the large springs described above, there are many smaller ones oozing from its banks. Three quarters of a mile below this set of springs, in the valley on the north side of the creek, rises another spring nearly as large as the largest one described above; water exactly the same. This splendid fountain of mineral water is situated at the north-east end of the town of Lampasas, and is the spring, on account of the accommodations afforded there, that is principally resorted to by the afflicted. Seven miles below this, in the valley on the south side of the creek, is still another quite flush spring of the same kind of water. Altogether, these several fountains of medicated waters, being precisely the same in their sensible properties, must evidently arise from the same reservoir. In the aggregate, according to the best estimate I could make, these springs discharge three hundred and sixty barrels of water, heavily charged with several mineral principles, every minute. Just think of it! Whence comes this inexhaustible supply of salt and sulphur, magnesia and iron, and perhaps several mineral principles not yet detected?

Previous to the war, in the proper season, invalids laboring under all forms of disease might be seen at these springs, many of whom were relieved of their complaints by the use of the waters, whilst others again received no benefit. The cases that seemed most favorably affected by its *proper* application were dyspepsia, indigestion, enlargement of the spleen, all skin diseases, sore eyes, mercurial and syphilitic taints, and other chronic complaints. I will here give it simply as my opinion, predicated on too small a number of cases to establish it, that the waters of these springs, carefully and systematically applied, after the manner of the water-cure doctors, would probably result more favorably in the removal of many forms of diseased action than would pure water alone, applied hydropathically. The few cases in which I applied it, after the manner of the water-cure doctors, consisted of enlargement of the spleen, abdominal dropsy, dyspepsia and a case or two of chronic sore eyes. The results were quite satisfactory.

There are several fresh-water springs gurgling up from the same locality, constituting this the most favorable point for a grand *water-cure* establishment that can be found in this or any other country, unless the celebrated Sour Lake waters may be excepted. Were I a young man and ambitious of fame and a cash fortune, I would go instantly to work at the Lampasas Springs. There would be no risk in building up a large water-cure establishment at that place; and I would greatly prefer the upper set of springs. Properly arranged in the best hydropathic style, I do not know of an enterprise that would attract so much attention, or that would remunerate the investment of capital so thoroughly.

On Wixson's creek, Gonzales county, is another set or series of mineral springs,[2] different in the character of their chemical properties from the waters of the Lampasas springs. They are situated in a kind of basin, a mile and half in diameter, the north-eastern side of which is cut and drained by the San Marcos river. These springs, flowing into the basin from various points at the base of the overhanging iron ore hills that bound it, are of several distinct characters, differing widely in their chemical properties. All in the same valley, and not far apart, there are three chalybeate springs, two white sulphur springs, two warm springs, two gas springs, and several fresh-water springs, all flush and gushing forth their clear, transparent waters in great abundance. This can be made into a splendid watering place at little cost.

This place has been for many years resorted to by the invalids of the surrounding country, and it is said that many bad and painful forms of disease have been relieved by the healing powers of the waters of these valuable springs.

Another quite popular (and many testify that it is deservedly so) watering-place is known by the unattractive title of *Sour Lake*.[3] This place is annually becoming more and more popular and has the reputation of curing many complaints. It is already a fashionable watering-place, much resorted to by the people of Houston and the lower countries during the sickly season.

The Piedmont springs,[4] of Grimes county, is a place of considerable notoriety, and having the best accommodations of any spring in Texas, is a quite fashionable resort during the summer.

There are many other mineral springs in the State—a great number of them—mostly impregnated with sulphur. I have not, however, examined them, and, of course, can not speak of the chemical properties of the waters that flow from them. Indeed, nearly every county in Texas has its mineral springs.

4. THE WATER-POWER OF TEXAS

It appears to me that, to people desirous of emigrating to Texas, a knowledge of the great and sometimes extraordinary water-power that is found in our State would be encouraging and useful to them in making up their conclusions. To know that, in the country they contemplate moving to, besides its many other decided advantages, it possesses also sufficient water-power to drive all the machinery for manufacturing purposes that the State will ever need, would be a cheering consideration to the emigrant.

I shall speak of Texas west and south of the Brazos, and in doing this I am not, nor do I intend to say that it is a preferable country to the eastern and northern portions of our State. It is a fine country, containing many superior facilities for making the industrious tillers of the ground prosperous and happy.

The first and the greatest, most abundant, and most available water-power in the State, or anywhere else, is at New Braunfels, Comal county. Here is water enough, and places and fall[s] enough to afford ample power and space to put in motion, on a large scale, all kinds of machinery for every species of manufacture yet discovered by man. I should fail were I to attempt a description of this unsurpassed water privilege. To present a satisfactory picture of this magnificent natural arrangement for the aid of scientific and manly mechanical effort would require more power than I possess. I look upon the Comal springs, and the rushing river that they form when united, and which dashes away in its precipitate course adown the slope to the mountain, as one amongst the wonders of the world. The place should have been named "Piedmont," for it is at the foot of the mountain that it displays its unlimited and truly wonderful powers.

I visited the Comal last May, and found considerable machinery already in action, and places were being prepared to erect more. As yet, however, there has not been appropriated the one twentieth of the water that flows from the springs. It is but a few miles from where these springs gush from the mountain sides to where they dash their clear, transparent waters into the Guadalupe river. The Guadalupe, too, has in its course many excellent mill seats and several quite suitable sites for manufacturing establishments. Already there are some very good mills on that river.

The next in order, as a superior power, is at the Marble Falls on Colorado river, Burnet county. This is very great; but as it has been described in the article on Texas marble, it needs not that I repeat it here.

The next in order of power, perhaps, are the San Antonio Springs. These flow from the foot of the mountain five miles above the city of San Antonio in Bexar county. They are all large springs, which in a short distance run together, forming a deep, swift-running river, that passes through the city, and is bridged in many places. Ditches have been cut, ages ago, and the water thrown into them two or three miles above the city. These ditches traverse the city in many places, which, with the clear, blue waters of the river passing down near its centre, supply the place with abundance of good running water. From the elevation of the water in the ditches at their lower extremities, it is clearly demonstrated that any amount of water-power can be accumulated by enlarging them. Twenty miles below the city there is a considerable fall in this river, in Wilson county, forming a splendid site for mills and machinery.

At Fort Inge, in Uvalde county, on the Rio Leona, is sufficient water-power for a large establishment. At the San Marcos Spring, which is but little inferior as to quantity of water to the best of them, there is a good corn and flouring mill in successful operation, and this mill does not employ the one hundredth part of the water that dashes swiftly by its wheel. The mill is placed on the bank of the river about half a mile below where this immense volume of water surges up from beneath the base of the mountain.

Immediately below the Upper Sulphur Springs, on Sulphur Fork of Lampasas river, Lampasas county, is a series of fine sites for machinery, and water plenty. One mile below this great spring is now in successful operation a good corn and wheat set of mills. At Fisher's Falls, on Colorado river, Llano county, is another place that affords great water privileges. The localities I have mentioned are all capable of driving the machinery of larger manufacturing establishments, and they are all situated on the upper border of the great farming prairie country that spreads out coastwise.

There are many smaller mill-seats to be found throughout the upper and more undulating region in all the wheat growing country. The level, coast-wise portion of the State, say from 200 to 250 miles from the sea-shore upward, has to be sure, many eligible and most beautiful rock-bottomed millsites; they are, however, rather inconvenient to water. In all that portion of Texas, steam is employed.

Appendix 5

Published Writings of Gideon Lincecum

AMERICAN NATURALIST

"The Tarantula Killers of Texas." 1 (1867): 137–41.
"The Scorpion of Texas." 1 (1867): 203–205.
"The Tarantula." 1 (1867): 409–11.
"Pouched Rat." 6 (1872): 24–25.
"The Gregarious Rat of Texas." 6 (1872): 487–89.
"The Opossum." 6 (1872): 555–57.
"Swamp Rabbit." 6 (1872): 771.
"Texas Field Mouse." 6 (1872): 772–73.
"Habits of a Species of Sorex," 7 (1873): 483–84.
"The Agricultural Ant." 8 (1874): 513–17.
"Sweet Scented Ants." 8 (1874): 564.
"Robber Ants." 8 (1874): 564.
"The Gossamer Spider." 8 (1874): 593–96.

THE AMERICAN SPORTSMAN

"The Wood Rat." 4 (January 24, 1874): 258.
"The Animals of Texas" [*Prognathus faceatus*]. 4 (February 28, 1874): 346.
"Personal Reminiscences of an Octogenarian." 4 (September 12, 1874): 374–75; (September 19): 390; (September 26): 406; 5 (October 3): 6; (October 10): 26–27; (October 17): 38; (October 24): 50–51; (October 31): 70; (November 7): 86–87; (November 14): 106–107; (November 21): 122–23, (November 28): 138–39, (December 5): 154; (December 12): 170–71; (December 19): 186–87; (December 26): 202–203; (January 2, 1875): 218; (January 9): 234; (January 16): 250.

JOURNAL OF THE LINNAEAN SOCIETY, ZOOLOGY

"Agricultural Ant of Texas." 6 (1862): 29–31.

PRACTICAL ENTOMOLOGIST

"The Texas Cabbage Bug." 1 (1866): 110.

PRAIRIE FARMER

"The Texas Cabbage Bug." 34 (1866): 152.

PROCEEDINGS OF THE ACADEMY OF NATURAL SCIENCES

"A Collection of Plants from Texas." 13 (1861): 98.
"On an Ant Battle Witnessed in Texas." 18 (1866): 4–6.
"On the Grapes of Texas." 18 (1866): 6–7.
"On the Small Black Erratic Ant of Texas." 18 (1866): 101–106.
"On the Agricultural Ant of Texas." 18 (1866): 323–31.
"On the Cutting Ants of Texas—*Oecodema texana buckley*." 19 (1867): 24–31.

PUBLICATIONS OF THE MISSISSIPPI HISTORICAL SOCIETY

"Autobiography of Gideon Lincecum." 8(1904): 443–519.
"Choctaw Traditions about Their Settlement in Mississippi and the Origin of
    Their Mounds." 8 (1904): 521–42.
"Life of Apushimataha." 9 (1906): 415–85.

THE SOUTHERN CULTIVATOR

"The Grasses of Texas." 19 (1861): 33–34, 51–52.

TEXAS ALMANAC

"Native or Indigenous Texas Grasses." 1861: 139–43.
"Botany—Directions by Which the Poisonous Plants of Texas May Be Readily
    Recognized." 1861: 143–44.
"The Cotton Worm." 1867: 195–96.
"The Indigenous Texian Grasses." 1868: 76–77.
"Gypsum in Texas." 1868: 85–86.
"Texas Marble." 1868: 87–88.
"Medicated Waters of Texas." 1868: 88–90.
"Waterpower of Texas." 1868: 90–91.

THE ZOOLOGIST

"The Cutting Ant of Texas." 3 (1868): 1270–81.

Notes

FOREWORD

1. Samuel Wood Geiser, *Naturalists of the Frontier*, p. 15.
2. *Webster's New International Dictionary Unabridged*, 2nd ed. (Springfield, Mass: G.&C. Merriam Co., 1955), p. 2238.
3. Geiser, *Naturalists of the Frontier*, p. 253–54.
4. Ibid., p. 271.
5. Ibid., p. 13.

INTRODUCTION

1. Letter to Dr. W. A. Dunn, June 19, 1861. Lincecum Papers, Center for American History, University of Texas at Austin. Unless otherwise indicated, all of Lincecum's letters cited in this volume are from this collection. We have modernized some of his punctuation for clarity but preserved the sometimes idiosyncratic spelling and grammar that flavor his writing. We have consistently italicized genus and species names, whether Lincecum underlined them or not.
2. Jerry Bryan Lincecum and Edward Hake Phillips, eds., *Adventures of a Frontier Naturalist: The Life and Times of Dr. Gideon Lincecum*, pp. xxi–xxxi.
3. There are thirteen boxes of Lincecum Papers in the Center for American History of the University of Texas in Austin, the bulk of them containing Gideon's correspondence in two versions. Lois Wood Burkhalter's assessment of these is apt: "The bulk of the letters have been typed and placed in as nearly chronological order as possible. The typed material is bound in fifteen volumes of almost 5,000 pages, frequently unnumbered. The typing, done in 1931 and 1932 by the National Youth Administration and through the courtesy of the Texas State Medical Association, is in many cases unreliable—names are wrong, Lincecum's handwriting has been misread, addresses are omitted. Copies should be checked against originals; and the

originals are difficult to locate." (*Gideon Lincecum, 1793–1874: A Biography*, p. 327). By the 1990s, the holograph originals, which were letterpress copies (the nineteenth-century equivalent of carbon copies), have deteriorated to the point where one is grateful that the typescripts, erroneous as they are, were made some sixty years ago, when the originals were more legible. Without the typescripts, some of the originals would now be undecipherable. Only when the original and the typescript of a given letter have been compared can the researcher feel confident of having come as near to Lincecum's original message as possible.

4. Burkhalter, *Gideon Lincecum*, p. 172.

5. Mason Locke "Parson" Weems (1759–1825) is best known for having originated the cherry tree myth about George Washington, but he was a significant popularizer of American history. Emily E. F. Skeel, "Mason Locke Weems," *Dictionary of American Biography*, Vol. 10, Pt. 1, 604–605.

6. Lincecum and Phillips, *Adventures of a Frontier Naturalist*, p. 36.

7. Ibid., p. 37.

8. Desmond King-Hele, *The Essential Writings of Erasmus Darwin*, pp. 11, 61–69, 206.

9. William Morton Wheeler, *Ants: Their Structure, Development and Behavior*, pp. 286–88.

10. Bert Hölldobler and Edward O. Wilson, *The Ants*, p. 610.

11. James Rodger Fleming, *Meteorology in America, 1800–1870*, p. xix.

12. Preface to "Chahta Tradition," Lincecum Papers, Box 2E365, Center for American History, University of Texas at Austin.

13. For Gideon's views on the Civil War, see Jerry Bryan Lincecum and Edward Hake Phillips, "Civil War Letters of Dr. Gideon Lincecum: 'I am out and out a secessionist.'"

14. Burkhalter, *Gideon Lincecum*, p. 187.

15. The Smithsonian has ten separate files of accessions from Gideon Lincecum dating 1867–74. These files indicate that his collections ranged all the way from the smallest insects to skins of mammals and birds as well as shells and fossils. His diary for 1867 indicates that he shipped one box of specimens weighing one hundred pounds. On another occasion in 1867 he sent 417 moths. He continued collecting specimens for the Smithsonian after relocating to Tuxpan, sending 1,000 butterflies.

16. Burkhalter, *Gideon Lincecum*, p. 218. Burkhalter used the titles *Flora of the United States* and *Structural Botany*. The former is a mistake for Chapman's *Flora of the Southern United States*, and Gray's *Structural Botany* was published as *Botanical Textbook* prior to the 1870s.

17. Several of these "orphan boys" were probably Gideon's grandsons, whose father, Capt. George W. Campbell, husband of Leonora Lincecum, had served in the Confederate Army and later drank himself to death. Burkhalter, *Gideon Lincecum*, p. 82.

18. Letter to Prof. Joseph Leidy, August 29, 1874.

19. *Proceedings of the Academy of Natural Sciences*, 19 (March 1867), 32.

20. Elliot Coues to Gideon Lincecum, March 22, 1874, Lincecum Papers.

21. Gideon's letter to Coues was not among the letterpress copies in the

Lincecum Papers, and our attempts to locate it were futile, despite inquiries at three repositories of Coues' papers: the National Archives, the Smithsonian, and the State Historical Society of Wisconsin.

22. Elliot Coues and Joel Asaph Allen, *Monograph of North American Rodentia*. For Gideon's contributions, see pp. 39, 264, 281, 335–36, 366, 500, 504, 539, 712, 716, 731, 1068, and 1074.

23. W. Conner Sorensen, *Brethren of the Net: American Entomology, 1840–1880*, pp. 150, 263–64.

CHAPTER 1

1. The "Formica" are Lincecum's ants. These organisms are in the subfamily Myrmicinae, order Hymenoptera. The agricultural ants are also called harvesting or stinging ants; they are in the genus *Pogonomyrmex*. The cutting ants, *Atta*, that Lincecum studied are also in the same subfamily. Hölldobler and Wilson, *The Ants*, pp. 12–15.

2. *Aristida stricta* is Lincecum's ant rice. The genus *Aristida* includes the triple-awned grasses. In McCook's article "The Agricultural Ants of Texas" he states that Lincecum's ant rice had been identified as *A. stricta*, a tall yellow needle grass. William M. Wheeler identified Lincecum's ant grass as either *A. stricta* or *A. oligantha*. Wheeler, *Ants*, p. 286. *A. oligantha* Mich. is commonly known as oldfield threeawn. Frank W. Gould, *Common Texas Grasses*, pp. 26–27.

3. *Celtis occidentalis* is the common and prolific hackberry. Ellwood S. Harrar and J. George Harrar, *Guide to Southern Trees*, pp. 249–51. *Viburnum dentatum*, called arrow-wood, is a shrub with small white flowers and can grow to heights of twenty feet. Clair A. Brown, *Louisiana Trees and Shrubs*, pp. 234–35. *Ilex vomitoria*, popularly known as yaupon, is a member of the holly family with red berries. Harrar and Harrar, *Guide to Southern Trees*, p. 448. *Xanthoxylum (Zanthoxylum) carolinianum* is also known as the tickle-tongue, toothache tree, or prickly ash. These shrubs are in brushy areas of the Edwards Plateau, lower parts of Plains Country, and north central Texas; they grow usually in calcareous soil. Donovan S. Correll and Marshall C. Johnston, *Manual of the Vascular Plants of Texas*. p. 910. The mustang grape is known as *Vitis mustangensis* Buckl. This grape grows along stream bottoms, thickets, fence rows, sandy slopes, and especially in disturbed grounds in the east half of Texas. Ibid., p. 1017.

4. *Argemone mexicana* L. is the Mexican poppy. Ibid., pp. 664–65.

5. *A Calendar of the Correspondence of Charles Darwin, 1821–1882*, eds. Frederick Burkhardt and Sydney Smith. Gideon's letters are recorded as No. 3035 (December 29, 1860) and No. 3082 (March 4, 1861).

6. Ibid., No. 3112: Charles Darwin to George Busk (April 5, 1861).

7. Sorensen, *Brethren of the Net*, pp. 197–98. Although Gideon is not one of the eighty-five entomologists cited by name in *The Descent of Man*, a list of "the habits and mental powers of worker-ants" (p. 513) includes a number of items discussed in his two letters to Darwin, such as "They make roads as well as tunnels under rivers. . . ." What is most striking in Darwin's discus-

sion of ant behavior here is his tendency to anthropomorphize in the same manner as Lincecum. Lincecum published several additional articles on ants. See appendix 5, Published Writings of Gideon Lincecum.

8. Burkhalter, *Gideon Lincecum*, p. 186. Durand also corresponded with Lincecum's youngest daughter, Sallie, and among the Lincecum Papers are three of his letters to her which indicate that "despite his age, Durand entertained romantic feelings toward Sallie." Ibid., p. 200n.

9. Sallie (Sarah Matilda) was Gideon's twenty-seven-year-old daughter, the eleventh of his thirteen children. She shared his great love of science, especially botany. For more on her biography, see appendix 1, Index of Notable Correspondents.

10. *Proceedings of the Academy of Natural Sciences*, 14 (May, 1861), 98. "The collection above referred to, of Texas plants, consists of 682 species or varieties, 540 of which are *Exogenous*, divided as follows: *Polypetalous*, 207; *Monopetalous*, 244; *Apetalous*, 84; *Gymnospermous*, 5; One hundred and forty are *Endogenous*, of which 106 belong to *Cyperaceoe* and *Gramineoe*. One *Equisetum* and one *Fern*."

11. Burkhalter, *Gideon Lincecum*, p. 187.

12. Ibid., p. 187n.

13. Gideon Lincecum, Writings on Ants, 1865, Collection 369, Unit No. 8-Microscopic Red ant, Myrmica? [sic], The Ewell Sale Stewart Library, The Academy of Natural Sciences, Philadelphia.

14. Burkhalter, *Gideon Lincecum*, pp. 219–20.

15. Gideon Lincecum, "On an Ant Battle Witnessed in Texas."

16. Gideon Lincecum, "On the Small Black Erratic Ant of Texas."

17. In Box 2E363 of the Lincecum Papers is a "Scribbling Diary for 1864," which Lincecum used for a variety of purposes (chiefly in 1867), using the first blank page, headed "Memoranda Brought From Last Year," to list his "Catalogue of Ants," as follows:

No. 1. Large, black, tree ant
2. Agricultural ant
3. Horticultural ant
4. Red-headed tree ant
5. Small black erratic ant
6. Servile, middle-sized, brown ant
7. Smallest black " [From here on Gideon used the ditto mark most of the time instead of repeating "ant."]
8. Very small red "
9. Brown-headed small "
10. Slow moving small brown "
11. Slow moving small black "
12. Slow brown concealment " ? [sic]
13. Quick large, bright red "
14. Little dark red, sand "
[There is no No. 15]
16. Small, blackish, strong odor "

17. Black, next to the least "
18. Light brown, smallest "
19. Small brown, very numerous "
20. Small brown Rock ant
21. Common woodlouse "
22. Large black, secretive, Rock "
23. Microscopic, smallest Red "
24. Large bright brown, odorous "
25. 5 sizes, under Rocks in prairie "
26. Small pale yellow "
27. Small black belligerent "
28. Large black-headed, thorax brown "
29. Half-sized, brown sand, slow & timid "
30. Head and abdomen large, quick, brown, sand "
31. Long slim, abdomen pointed, timid "
32. Very smallest, pale shiny brown "
33. Small, brown head, abdomen semitransp[arent]
34. 2 sizes, small, brow[n], hairy, Rock "
35. Long, slim, redish brown, eyes large "
36. Small, long, slim black "
37. Large dark brown "
38. Large black, 2 sizes, transient, has no home "
39. Large black, head small
40. Head large, mouth and abdomen black "
41. Dark brown, small "
42. Size larger than No. 5, have red head, giant
43. Dark brown, large thorax, has a ja[u]nty appearance "
44. Wild, timid, brown, under dry logs "
45. Brown, have giant slaves, head & jaws large
46. Very rare, females large, Abdomen banded
47. Diminutive, und[er] old rails & rocks in prair[ie]
48. Black, slim, wild, Dwells in the [page torn]

18. Gideon Lincecum, "On the Agricultural Ant of Texas."
19. John Traherne Moggridge (1842–74) was a Fellow of the Linnean Society. He published his study, *Harvesting Ants and Trapdoor Spiders: Notes and Observations on Their Habits and Dwellings*, in 1873. A supplement was published in 1874. *National Union Catalogue*. He studied harvester ants in Southern France and "established that harvesters play an important role in dispersing plants by accidentally abandoning viable seeds in the nest vicinity or failing to deactivate them before they sprout." Hölldobler and Wilson, *The Ants*, p. 609.
20. Auguste Forel (1848–1931) was a Swiss psychiatrist and entomologist who published his *Ants of Switzerland* in 1875. Burkhalter, *Gideon Lincecum*, p. 203n.
21. Henry Christopher McCook, *The Natural History of the Agricultural Ant of Texas*, pp. 11–13.

22. Burkhalter has a good treatment of the controversy over Lincecum's anthropomorphic interpretation of ant behavior and leans in his favor. Burkhalter, *Gideon Lincecum*, pp. 219–26.

23. William Morton Wheeler (1865–1937) was a prodigious student of entomology who served the last thirty years of his distinguished career at Harvard's Biological Laboratories. He published his monumental *Ants: Their Structure, Development, and Behavior* in 1910. In 1923 he issued *Social Life Among the Insects*, followed in 1928 by *The Social Insects: Their Origin and Evolution*, titles at least that would have appealed to Lincecum. F. M. Carpenter, "William Morton Wheeler," *Dictionary of American Biography*, Vol. 11, Pt. 2, 707–708.

24. Wheeler, *Ants*, p. 286.

25. Ibid., pp. 187–89.

CHAPTER 2

1. Lincecum and Phillips, *Adventures of a Frontier Naturalist*, pp. 135–39. For a thorough treatment of Lincecum's relationship with the Choctaws in Mississippi, see Cheri Lynne Wolfe, "The Traditional History of the *Chahta* People: An Analysis of Gideon Lincecum's 19th Century Narrative." Wolfe edited and annotated Lincecum's 650-page manuscript of an oral tradition that he took down from a Choctaw sage over a four-year period in the early 1820s. It is considered the most extensive archive of Southeastern Indian lore collected before 1830.

2. Ibid., p. 144.

3. Probably Lincecum was referring to the early American botanist Benjamin Smith Barton (1766–1815), whose *Elements of Botany* (1803) was a pioneering effort. Harris E. Starr, "Benjamin Smith Barton," *Dictionary of American Biography*, Vol. 1, Pt. 2, 17–18.

4. Gideon Lincecum, "Botany—Directions by which the poisonous Plants of Texas may be Readily Recognized." Among the Lincecum Papers is Gideon's herbarium. Boxes 2E367, 2E368, and 2E369 contain botanical specimens which are organized using the classification system of Amos Eaton, as set forth in *Manual of Botany for North America*, published in 1829.

5. Burkhalter, *Gideon Lincecum*, pp. 173 and 218. John Darby's *Botany of the Southern States* was published in 1855, Alvan W. Chapman's *The Flora of the Southern United States* in 1860.

6. In the context of Lincecum's writing, the term "morus multicaulis" may refer to the mulberry tree. The genus for the mulberry is *Morus*; however, the species *multicaulis* is no longer used. The white mulberry is *M. alba*, and was introduced to the eastern United States from China. Correll and Johnston, *Manual of Vascular Plants of Texas*, pp. 496–97. The widespread planting of mulberry trees in the eastern United States in the early nineteenth century was a classic case of introducing foreign plants to an area rather than emphasizing natives.

7. Gideon Lincecum, "Native or Indigenous Texas Grasses." Thirteen distinct species of Texas grasses are described. See appendix 3 for an excerpt from

this article. Another version of this article appeared in *Southern Cultivator* as "Grasses of Texas, Part 1" and "Grasses of Texas, Part 2."

8. Rescuegrass, *Bromus unioloides* (Willd.) H. B. K., was introduced into the western United States from South America as a forage grass but is now growing mostly as a weed of ditches, vacant lots, old fields, and roadsides. It is not to be confused with Tall fescue or *Festuca arundinacea* Schreb., which was introduced from Europe. It is a cool-season perennial that is widely established in temperate and cool regions of North America. In Texas it is significant mainly as an improved pasture grass. In our region it persists under natural conditions. Gould, *Common Texas Grasses*, pp. 69, 129–30.

9. *Agrostis* is also called winter bent grass or winter redtop. It is a cool-season weak perennial of only fair forage value for livestock and wildlife but is good for birdseed. Gould, *Common Texas Grasses*, pp. 12–13. There are three possible plant species Lincecum may be referring to as *Stipa*: *S. avenacea* (blackseed needle grass), *S. comata* (needle grass), and *S. leucotrichia* (Texas winter grass or spear grass). Correll and Johnston, *Manual of Vascular Plants of Texas*, pp. 121–23. *Cynodon dactylon* (L.) Pers. is Bermuda grass and provides good grazing for livestock and wildlife. Gould, *Common Texas Grasses*, pp. 86–87.

10. Brome grass is the genus *Bromus*, and there are about 150 species in the temperate regions. *Bromus carinatus* is a coarse, tufted, short-lived perennial grass. Henry A. Gleason and A. Cronquist, *Manual of Vascular Plants of Northeastern United States*, pp. 770–71. *Bromus ciliatus* carries the name of *B.Richardsonii* Link today; it grows in moist woods and banks. *Gray's New Manual of Botany*, pp. 162–65. Correll and Johnston, *Manual of Vascular Plants of Texas*, pp. 123–26. Canary grass is classified in the genus *Phalaris*. Lincecum terms this grass *P. intermedia*; today it is known as *P. caroliniana* Walt. It grows in all regions of Texas in grasslands and in open woodlands. Gould, *Common Texas Grasses*, pp. 190–92. White clover is now classified as *Trifolium repens* L. Perhaps this is the clover Lincecum was referring to. Correll and Johnston, *Manual of Vascular Plants of Texas*, pp. 806–809.

11. Austin grass or *Panicum obtusum* H. B. K. was what Lincecum called *P. gibbum*. Today it is also called vinemesquite. It is a native, warm-season perennial that provides good grazing for livestock. Gould, *Common Texas Grasses*, pp. 171–73.

12. Gideon Lincecum, "The Indigenous Texas Grasses."

13. *Vitis lincecumii* Buckl. is the post oak grape; S. B. Buckley named this grape for Gideon. The mustang grape is properly known as *V. mustangensis* Buckl. Correll and Johnston, *Manual of Vascular Plants of Texas*, pp. 1015–21.

14. Gideon Lincecum, "Grapes of Texas."

15. Today the post oak is classified as *Quercus stellata* Wang. It is one of the dominant species in the Cross Timbers in central Texas. Correll and Johnston, *Manual of Vascular Plants of Texas*, pp. 475–76.

16. The black oak is *Quercus velutina* Lam. Ibid., p. 491.

17. The pecan, *Carya olivaeformis*, is now *C. illinoensis*. Ibid., p. 460. Lincecum's "Spanish buckeye" might be the genus *Ungnadia* Endl., also known as the

Texas buckeye or the Mexican buckeye. Its family is the Sapindaceae or soap-berry family. The seeds are spherical, 1 to 1.5 cm in diameter, dark brown to blackish, and poisonous. Ibid., pp. 1005–1006. Another possibility would be the genus *Aesculus* L., the buckeye or horse-chestnut, which is in the family Hippocastanaceae or Buckeye family. Ibid., pp. 1004–1005. The elms are in the genus *Ulmus*, and *U. americana* L. is the American elm; the red elm is *U. rubra* Muhl. Ibid., pp. 494–95. The wild peach or laurel cherry was known as *Cerasus caroliniana* to Lincecum; however, today we call it *Prunus caroliniana*. J. C. Willis, *A Dictionary of the Flowering Plants and Ferns*, p. 225. *Algarobia* is now called *Prosopis. P. glandulosa* is the honey mesquite tree. Ibid., p. 40.

CHAPTER 3

1. The grasshoppers belong to the class Insecta, the order Orthoptera, and the family Locustidae. Donald J. Borror and Dwight M. DeLong, *An Introduction to the Study of Insects*, pp. 123–24, 130–32.
2. Gideon refers to "genus Locusta" but he probably means the grasshoppers in general. The Family is named Locustidae.
3. Cicada are members of the family Cicadidae and there are some 200 species in North America. Borror and DeLong, *Study of Insects*, pp. 250, 252.
4. The term "ichneumon" probably refers to the Ichneumon wasps (order Hymenoptera, family Ichneumonidae), which are parasitic on caterpillars or other insect larvae. The term "Libellula" may come from the family Libellulidae (order Odonata), which includes the dragonflies. "Epizooty" refers to a disease which is prevalent temporarily among animals.
5. Order Coleoptera, the beetles, contains about 40 percent of the known species in the class Insecta. Borror and DeLong, *Study of Insects*, p. 296.
6. The term "alum" refers to a crystalline substance composed of a double sulfate of aluminum and potassium which is used in medicine and dyeing. Clayton L. Thomas, *Taber's Cyclopedic Medical Dictionary*, p. 69.
7. The red-dyed grasshopper that Lincecum refers to is a member of the family Acrididae, subfamily Cyrtacanthacridinae, commonly known as the spur-throated grasshoppers. The most injurious grasshoppers belong in this group. A possible identification of this red-legged grasshopper would be *Melanoplus femur-rubrum* (DeGeer). This grasshopper could turn the alcohol solution red when the dye was extracted out of it. Borror and DeLong, *Study of Insects*, pp. 130–32.
8. The Jerusalem oak is an annual weed of dry, sandy, gravelly areas. It is in the goosefoot family, and its scientific name is *Chenopodium botrys* L. The Great Plains Flora Association, *Flora of the Great Plains*, pp. 160, 169.
9. Honeybees are members of the order Hymenoptera, family Apidae. Borror and DeLong, *Study of Insects*, p. 657.
10. François Huber (1750–1831) was a Swiss naturalist whose main specialty was the study of bees. His pioneering work, *Nouvelle observations sur les abeilles*, 1792, was translated into English in 1806 and updated in 1814. "Francois Huber," *Encyclopedia Americana*, 14:465.

11. The bee moth is *Galleria mellonella* L. The female lays her eggs at night in the hives of honey bees. The developing larvae feed at night to prevent being stung and thrown out of the hive. Poorly kept hives are seriously damaged by larval webbing and destruction of combs. Ross H. Arnett, *American Insects*, p. 570.

12. Spiders are members of the class Arachnida, order Araneae. Robert B. Barnes, *Invertebrate Zoology*, p. 619.

13. *Mygale Hentzii* is Lincecum's name for the tarantula. Tarantulas are members of the suborder Opisthothelae (mygalomorphs). Ibid., p. 637. The species name *Hentzii* refers to Nicholas Marcellus Hentz (1797–1856), best remembered for his descriptions of 141 species of spiders, which were published in the journal of the Boston Society of Natural History between 1842 and 1850. These provided the foundation upon which knowledge of North American spiders is based. Hentz's name was given to a dozen spider species, including a common tarantula. John Cooke, "A Pioneering Spider Man." Lincecum has an article entitled "The Tarantula," which appeared in the *American Naturalist* in 1867.

14. The "trap-door" spiders are members of the suborder Opisthothelae (mygalomorphs). Barnes, *Invertebrate Zoology*, p. 619.

15. Lincecum's "balloonists" are spiders which are lifted up by the wind currents and can be blown some distance away. In this way the distribution of spiders is promoted. Willis J. Gertsch, *American Spiders*, pp. 28–30.

16. The dirt daubers or mud daubers are solitary wasps of the family Sphecidae. Borror and DeLong, *Study of Insects*, p. 737.

17. Burkhalter identifies Dr. A. G. Lane as an old ministerial friend of Lincecum's from Mississippi who had settled in Lockhart, Texas. Burkhalter, *Gideon Lincecum*, p. 106.

18. "Pompilus" refers to wasps in the family Pompilidae, the spider wasps. These organisms are slender, with long spiny legs. They capture and paralyze a spider and then place it in a cell that they have prepared for it. Borror and DeLong, *Study of Insects*, p. 736.

19. Gideon discusses the wasp, *Pompilus formosus*, in the article "The Tarantula Killers of Texas" in the *American Naturalist*.

20. The term "scarabacus" probably refers to the scarab beetles which feed either on dung or plant material. They are in the family Scarabaeidae. Borror and DeLong, *Study of Insects*, p. 385.

21. The flat paper nests are made by wasps of the family Vespinae, subfamily Polistinae, or the social vespids or paper wasps. Arnett, *American Insects*, pp. 452–54.

22. The "negus tree" has proven elusive. One possibility is the *Acer Negundo* L. or box elder tree. Correll and Johnston, pp. 1001–1002.

23. The yellow jacket wasps are social vespids of the genus *Vespula*. Borror and DeLong, *Study of Insects*, p. 731.

24. Centipedes are Myriapodous arthropods in the class Chilopoda. They live in soil or humus, beneath stones, bark, and logs. Barnes, *Invertebrate Zoology*, pp. 810–11.

25. *Tephroid virg.* refers to the organism *Tephrosia virginiana*, or goat's rice. This plant is a member of the legume family (Fabaceae). The roots contain rotenone, an insecticide and fish poison. In a weakened form, the material was used by the Native Americans and early settlers as a medicine. Geyata Ajilvsgi, *Wildflowers of Texas*, p. 187.

26. *T. heros* is referring to a centipede. One of the genus names in the order Scolopendromorpha is *Theatops*. Perhaps that is what the "T." refers to. Barnes, *Invertebrate Zoology*, pp. 816–17.

27. Scolopendra refers to the order Scolopendromorpha. Ibid., pp. 816–17.

28. *S. heros* may well mean a centipede in the order Scolopendromorpha. Ibid., pp. 816–17.

29. Gideon is not referring to centipedes as reptiles but rather is using the term "reptile" to mean "moving on the belly or by means of small, short legs."

30. Gideon's "spider that limps along" might be a jumping spider in the order Salticidae. These spiders are active in the daytime and walk with an irregular gait. Barnes, *Invertebrate Zoology*, pp. 639–42.

31. Probably Lincecum is referring to David Tidwell of Union Hill, Washington County, who is listed in the U.S. Census of Texas in 1860, p. 160.

32. Despite what Lincecum may have thought, scorpions are in the class Arachnida, order Scorpiones. What may have confused him is the fact that development in scorpions is either ovoviviparous or truly viviparous. At birth, the young are only a few millimeters long, and they immediately crawl upon the mother's back. The young remain there through the first molt (about one week). Then they gradually leave the mother and become independent. They reach adult stage in about one year. Barnes, *Invertebrate Zoology*, pp. 608–609. Lincecum published an article entitled "Scorpions of Texas" in the *American Naturalist* in 1867.

33. Based on Lincecum's description of the daddies and the "spicy" odor they produce, he is probably describing the daddy-long-legs which are members of the order Opiliones. They are abundant in vegetation, forest floors, tree trunks, fallen logs, and humus. They have a pair of scent glands which produce a secretion composed of quinones and phenols; this secretion has an acrid odor. Some of the daddies spray their prey while others pick up a droplet of secretion and mix it with regurgitated gut fluid and thrust it at the would-be predator. A second identification could be a group of arthropods known as the Pholcidae; these have long, slender legs with flexible ends and are white to gray in color. They make webs in dark corners of houses and cellars. Barnes, *Invertebrate Zoology*, pp. 639, 642–43.

34. This is probably a beetle in the family Hydrophilidae, the water scavenger beetles. Ross H. Arnett, *The Beetles of the United States*, pp. 215–25.

35. Judging from the context, Lincecum's "green Boat flies" may well be beetles. There are minute-sized beetles in the water scavenger family. Lester A. Swann and Charles S. Papp, *The Common Insects of North America*, p. 353.

36. The "clever little monthly" to which Lincecum is referring was the *Practical*

Entomologist, which had a short life, 1865–67. Cresson was its editor 1865–66, and Benjamin D. Walsh was the editor 1866–67. Sorensen, *Brethren of the Net*, pp. 75, 320.

37. Lincecum's cabbage bug was the subject of a paper, "The Texas Cabbage-Bug" in the *Practical Entomologist*. Lincecum sent specimens of these insects to the editor. Before the beginning of the article, the editor has a short paragraph identifying the organism as a true bug in the family Scutelleridae, order Heteroptera (now Homoptera), and named it *Strachia histrionica* Hahn.

38. Gideon Lincecum, "The Cotton Worm."

39. Lincecum continued to collect butterflies after moving to Mexico in 1868 and sent 1,000 specimens to the Smithsonian from there. Burkhalter, *Gideon Lincecum*, p. 257.

40. The term "Libubilia" is perhaps a mistake for Libellulidae (order Odonata), which includes the dragonflies.

41. The "Cow-Killer" is a member of the family Mutillidae, or the velvet ants. This organism is brick red and black and covered with thick plush-like hairs. Its length is 16–30 mm. The males have wings, but the females lack them and look "ant-like." The females seem innocent but sting viciously if picked up. Borror and DeLong, *Study of Insects*, p. 723.

42. The cow-killer and ant lion are not the same organism. The ant lion is in the family Myrmeleontidae. An ant lion adult is similar in general appearance to the damselflies, with long, multiple-veined wings and a long, slender abdomen; however, there are differences, such as a soft body, clubbed antennae, and wing-venation patterns. Their larvae are known commonly as doodle-bugs. Ibid., p. 293.

43. "Aptera" was part of an old classification scheme and included the various wingless arthropods.

44. Most likely, Lincecum is referring to the work of Thaddeus William Harris (1795–1856), the Massachusetts physician and entomologist who became librarian at Harvard from 1831 to 1856. His *Report on the Insects of Massachusetts, Injurious to Vegetation*, published in 1841, was a truly pioneering work and had "an enormous influence." Sorensen, *Brethren of the Net*, p. 70. A second enlarged edition, with many engravings, was published after his death, edited by Charles L. Flint (1824–89), not by Asa Fitch, as Lincecum (or Buckley) mistakenly stated. Harris authored no book with the title *On Entomology*, though one volume of excerpts (1845) bore the title *On Insects*. Ibid., pp. 11, 70. Leland O. Howard, "Thaddeus William Harris," *Dictionary of American Biography*, 4, Pt. 2, 321–22. Asa Fitch (1809–79), a gentleman farmer and naturalist living near Salem, New York, became the first full-time professional entomologist in the United States when he was appointed in 1854 to a post created by the New York state legislature. Between 1855 and 1872 he issued fourteen annual reports on applied entomology, designed to aid farmers in dealing with insect pests, but he did not edit Harris's great work. Ibid., pp. 71–72.

CHAPTER 4

1. For Lincecum's bird encounters, see Lincecum and Phillips, *Adventures of a Frontier Naturalist*, pp. 152–53, 164, 168–70, and 195.
2. Gideon Lincecum, "Meteorological Journal for 1860," Lincecum Papers, Box 2E363.
3. The white-fronted goose is *Anser albifrons*. Chandler S. Robbins, Bertel Bruun, and Herbert S. Zim, *A Guide to Field Identification of Birds of North America*, pp. 42–43. Probably the swans were the tundra swans (also called whistling swans) *Cygnus columbianus*. Roger T. Peterson, *A Field Guide to Western Birds*, pp. 16–17.
4. The chapparal cock or roadrunner is a member of the cuckoo family (*Cuculidae*), but its Latin designation is *Geococcyx californianus*. Peterson, *Field Guide*, pp. 85–86.
5. What Lincecum referred to as *Q. catesby* (*Catesbaei*) or the turkey oak, is now called *Q. laevis*. Thomas Elias, *Complete Trees of North America*, pp. 364–65.
6. The swallow-tailed hawk is probably the swallow-tailed kite, *Elanoides fortficatus*. Robbins, Bruun, and Zim, *Guide to Field Identification*, pp. 66–67.
7. Lincecum's three-month field trip to West Texas is covered well in Burkhalter, *Gideon Lincecum*, pp. 230–37. He kept a daily diary on the trip which is in Box 2E363, Lincecum Papers. It is included in the "Scribbling Diary for 1864," which Lincecum modified to serve as a diary for 1867. An entry for June 17, 1867 (after his return to Long Point), reads: "wrote a letter to the Governor, or rather finished one to him, giving a short account of the mineral resources of Texas, examined by me during my travel of three months in 18 counties. Washington, Burleson, Bastrop, Travis, Williamson, Milam, Falls, Bell, Coryell, Lampasas, Burnet, Blanco, Hays, Comal, Guadeloupe, Gonzales, Caldwell, Gillespie, and Bexar." Counting his home county of Washington, the list numbers 19.
8. According to ornithologist Karl Haller, of Austin College, Lincecum's bird skinning technique would provide a functional study skin for collections. Lincecum was well aware that the brain tissue in the skull needed to be removed but was not able to accomplish the task.
9. Based on Lincecum's description of the behavior of a very small grass sparrow which runs along like a mouse, this bird may be LeConte's sparrow (*Passerherbulus caudacutus*). Robbins, Bruun, and Zim, *Guide to Field Identification*, pp. 310–11. The Longspurs are in the genus *Calcarius*. *Field Guide to Birds of North America*, pp. 408–11. Lincecum's use of "Ploctrophanes" might be the word "Plectrophenax." Today, this word is used for the genus of the buntings. A "pretty sparrow with rusty but[t]s to his wings" is probably referring to the vesper sparrow, *Pooecetes gramineus*. Robbins, Bruun, and Zim, *Guide to Field Identification*, pp. 312, 324.
10. Lincecum's diary for 1867 has numerous references to Dr. Ruff as a physician in Long Point. In an 1869 accession file at the Smithsonian, George W. Lincecum, Gideon's grandson, refers to a bird specimen he is sending as coming from "Dr. D. E. Ruff."
11. The burrowing owl is *Speotyto cunicularia*, and the prairie owl may be the great horned owl, *Bubo virginianus*. The eagle-shaped hawk is probably a

ferruginous hawk or *Buteo regalis*. The butcher bird is probably the logger-head shrike or *Lanius ludovicianus*. The gnatbird or gnat catcher is in the genus *Polioptila*. The yellow-bellied sapsucker is *Sphyrapicus varius*. The snow bird that Lincecum is referring to is probably a dark-eyed junco, *Junco hyemalis* or perhaps is a snow bunting, *Plectrophaenax nivalis*. Robbins, Bruun, and Zim, *Guide to Field Identification*, pp. 70, 160–64, 184, 242, 236, 324.

12. The whooping crane is *Grus americana*. Ibid., p. 100.

13. The common snipes are known as *Capella gallinago*. The broad-billed, flat-footed, short-legged duck that Lincecum refers to could be a female wood duck, *Aix sponsa*. When she takes off from the water, she squeals. Another possible identification is the ruddy duck, *Oxyura jamaicensis*. Ibid., pp. 50–51; 60–61, 126. "San" is Gideon's son Lysander.

14. The curlews are in the genus *Numenius*. Lincecum might be referring to the long-billed curlew (*N. americanus*) or perhaps the eskimo curlew (*N. phaeopus*). The plovers are in the family Charadriidae. Ibid., pp. 110–15.

15. Lincecum is referring to his impending move from Texas to the new colony of ex-Confederates in Tuxpan, Veracruz State, Mexico. See chapter 8.

16. Baird's *The Birds of North America* (New York: Appleton and Co., 1860) was "an improved edition of the author's 'Birds,' which was issued in 1858 by the War Department as Vol. 9 of the Reports of explorations and surveys, to ascertain the most productive route for a railroad . . . to the Pacific Ocean" as described in the *National Union Catalogue*. Probably this was "the catalogue of N. A. birds" which Lincecum recently received from Baird. Letter to Spencer Baird, April 1, 1868.

17. Gideon's grandson, George W. Lincecum, was the great-grandfather of the co-editor of this volume, Jerry Bryan Lincecum. Records obtained from the Smithsonian indicate that he alone among Gideon's descendants continued the collecting that his grandfather had pursued so assiduously just prior to his departure for Tuxpan in 1868. Four accession files include letters from G. W. to Spencer Baird in 1868–69, indicating that Baird was generous in encouraging the efforts of this young man. George wrote that he had studied Gideon's books and letters in order to prepare himself to become a naturalist. Like his grandfather, he provided the Smithsonian with good descriptions of the behavior of the birds and small animals whose skins he sent. There are no records of any contact after 1869, however.

18. For Lincecum's youthful hunting exploits, see Lincecum and Phillips, *Adventures of a Frontier Naturalist*, pp. 20–24.

19. Solomon B. Cox of Belton, Texas, had been a member of several volunteer Ranger companies under John Henry Brown in Bell County in 1859 and 1860, protecting the frontier against raids by Native Americans. George W. Tyler, *The History of Bell County*, pp. 187, 189–90.

20. Lincecum's account of the "cannibals" is published as "Habits of a Species of Sorex" in *The American Naturalist*. The genus *Sorex* (family Soricidae) is the long-tailed shrew. Shrews are small, mouse-like animals with long, pointed noses. These animals are basically solitary and are highly aggressive toward one another. If several are kept together in close quarters, only one

individual may survive. Ronald M. Nowak, *Walker's Mammals of the World*, pp. 145–49.

21. For a complete listing of Lincecum's publications, see appendix 5, Published Writings of Gideon Lincecum.

22. Lincecum published two articles in the *American Naturalist* in 1872: "The Opossum" and "The Gregarious Rat of Texas *(Sigmodon Berlandierii)*."

23. The "genus *Talpa*" refers to organisms in the family Talpidae, the moles, shrew moles, and desmans. These animals burrow extensively, spending much of their lives underground, and there are some aquatic or semi-aquatic forms that occasionally burrow. Currently, the genus *Talpa* is reserved for the old-world moles. North American moles that Lincecum might be referring to are *Scalopus aquaticus* (the eastern American mole), *Parascalops breweri* (the hairy-tailed mole), and *Condylura cristata* (the star-nosed mole). Nowak, *Walker's Mammals of the World*, pp. 167–79. Lincecum used the name "salamander" to refer to the plains pocket gopher, *Geomys bursarius*, and Elliott Coues noted that this was a common practice in some regions. Probably the term "salamander" was a corruption of "sand mounder." William B. Davis, *The Mammals of Texas*, pp. 145–47.

24. The "horned frog" is, as Lincecum stated, a lizard, *Phrynosoma*. This organism has spiky scales that give it a grotesque appearance. Also, it has a placid temperament and makes a harmless and interesting pet. Cleveland P. Hickman, Sr., Cleveland P. Hickman, Jr., and Frances M. Hickman, *Integrated Principles of Zoology*, p. 512. In a letter to Spencer Baird dated August 19, 1868, George W. Lincecum writes: "Have you ever received any Horned Frogs from Texas? A very curious animal indeed, will live half a year shut up in a bottle. If not, I will send you some."

25. Cope was indeed a Quaker; he attended Friends School and the University of Pennsylvania. See appendix 1, Index of Notable Correspondents, for more on his biography.

26. Sarah Bryan Lincecum (1797–1867) was born and raised in Georgia and married Gideon on October 25, 1814. She bore him thirteen children and tolerated his various scientific interests and idiosyncracies. Burkhalter, *Gideon Lincecum*, pp. 69, 229.

27. "John H. Douglas of Mason City, Illinois" has eluded our search.

28. Lincecum has a fossil named after him. *Pholadomya lincecumi* Shumard is an elongated species of clam from the upper Cretaceous, Taylor group. It has prominent radiating ridges. Charles Finsley, *A Field Guide to Fossils of Texas*, p. 77, plate 65.

29. William Spillman is listed as residing in Columbus in the 1850 U.S. Census for Lowndes County, Mississippi, p. 123.

CHAPTER 5

1. It is somewhat an exaggeration to refer to the hills of Travis, Burnet, Llano, and Blanco counties as "mountains," but many of the hills have been so

designated by local usage, perhaps because of their contrast to the flatness of so much of the Texas terrain.

2. These Marble Falls are on the Colorado River in Burnet County. A dam was constructed above the falls in 1949–51, creating Marble Falls Lake and providing for the production of hydroelectric power at the site. The town of Marble Falls was not founded until 1887. A textile factory was built at the falls in the 1890s but was only moderately successful. The town became an important trade center, including trade in wool and mohair, but its population was only 2,161 in 1960. "Marble Falls Lake," *Handbook of Texas*, 3:573; "Marble Falls, Texas," *New Handbook of Texas*, 4:500.

3. Lincecum had lived on or near the Tombecbee (Tombigbee) River in Mississippi from 1818 to 1848.

4. Yegua Creek flowed eastwardly toward the Brazos River and formed the northern boundary of Lincecum's Washington County. He lived five miles south of the creek at Long Point. This part of the creek is now under Somerville Lake. "Yegua Creek," *Handbook of Texas*, 2:943; "1976 Annual Report of the Chief of Engineers on Civil Works Activities," Department of the Army, 2:xvi, 17.

5. Barton Creek flows into the Colorado River on the south side of the city of Austin. It is named for "Uncle Billy" Barton, one of the earliest pioneers of the Austin area. "William Barton," *Handbook of Texas*, 1:118; *New Handbook of Texas*, 1:404.

6. Burkhalter, *Gideon Lincecum*, p. 176.

7. Dr. Francis Moore, Jr. (1808–64), born and raised in New England, migrated to Texas in 1836 and settled in Houston, where he was a newspaper editor and mayor. He became interested in geology and served as state geologist, 1860–61. He returned to the North when the Civil War broke out. The so-called "Iron Mountain" in Llano County turned out to be a myth, according to Samuel Wood Geiser. "Francis Moore, Jr.," *New Handbook of Texas*, 4:817–18; Burkhalter, *Gideon Lincecum*, p. 185.

8. Dr. Benjamin Franklin Shumard (1820–69) was another Northern scientist who came to Texas before the Civil War and served as Texas State Geologist, 1858–60. See appendix 1, Index of Notable Correspondents, for more information.

9. See appendix 1, Index of Notable Correspondents, for notes on Governor Throckmorton.

10. Burkhalter says Lincecum's party explored twenty-eight Texas counties on this expedition, but Lincecum lists only nineteen. Apparently Burkhalter added the counties Lincecum covered in an East Texas expedition later in the year. Burkhalter, *Gideon Lincecum*, p. 237.

11. Lincecum's diary for 1867 notes that on May 31 he called on Gov. Throckmorton, who asked him to write a letter reporting on his expedition. An entry for June 13 indicates that he finished writing the letter and was preparing to mail it when he heard a rumor that the governor had been removed from office. However, this proved false; Throckmorton was not

removed until July 30, 1867. Lincecum sent him the letter and it was well received. Burkhalter, *Gideon Lincecum*, pp. 237–38.

12. Mormon Mills was the site of a Mormon settlement founded by Lyman Wight in 1851, which Wight abandoned in 1853. "Mormon Mills Colony," *Handbook of Texas*, 2:235; "Mormon Mill [*sic*] Colony," *New Handbook of Texas*, 4:840–41.

CHAPTER 6

1. "Gideon Lincecum's Journal of a Trip to Texas, 1835," in Burkhalter, *Gideon Lincecum*, p. 309.

2. Jacob Kuechler (1823–93), a native of Germany and well educated, had immigrated to Texas and settled at Fredericksburg, where he was a farmer and surveyor. Unlike Lincecum, he opposed secession and escaped to Mexico after the battle of the Nueces in 1862. He returned to Texas after the Civil War. After first appearing in a German-language newspaper in 1859, Kuechler's work on drought cycles was reprinted in the *Texas Almanac* for 1861. "Jacob Kuechler," *New Handbook of Texas*, 3:1164–65; Burkhalter, *Gideon Lincecum*, p. 129n.

3. The pin oak, or *Quercus palustris* Muench., is also called the swamp spanish or pin oak according to *Gray's New Manual of Botany*, p. 342.

4. The ozone paper that Lincecum refers to is described in two of his letters (Swante Palm, October 20, 1859; and Col. C. G. Forshey, March 29, 1860). To Palm, Gideon wrote, ". . . to determime the amount of ozone requires but very small slips for [the] experiment—one eighth of an inch wide and 2 inches long, is plenty—Bend a pin into a fish hook, attach it to a thread, hang it up in your little porch, in the free air, but out of the way of the rain, and upon this pin, every morning hook a little slip of the paper. It will exhibit the amount of ozone the next morning. I send also an ozone scale; I had no proper coloring matter, but this scale will serve to show the principle." To Forshey, Lincecum wrote, ". . . The Ozone paper, that I have prepared is always more or less colored when I take it down, sometimes as high as 7.0 on the ozone scale. I will enclose some ozone paper in this letter and you can try it. The highest coloring recorded on my register occured on the 18 inst. In the course of that 24 hours, the wind changed from N. to S. in the forenoon, then back by 9 P.M. to S. E. Generally the Norther is charged with the most Ozone, and it does not seem to make any difference about its humidity. When the wind is from the south, & the fog is heavy as to cause everything to drip, the Ozonometer rarely indicates more than 2. I have taken the paper down when it was perfectly saturated with water, and the Ozone would show but 2. On the other hand, when the paper would be dry, sky clear, wind N. and pretty strong Ozone 5. or 6."

5. On the folklore relating to the Texas norther, see Edward Hake Phillips, "The Texas Norther," *The Rice Institute Pamphlet*, 41, no. 4 (Jan. 1955), chapter 9: "The Norther as a Folk Theme," pp. 135–42.

6. In November, after the well had reached a depth of several hundred feet, Gideon suspended the project, as his workers were needed at other tasks, and the onset of the Civil War put an end to the experiment. Letter to Nephew John (Lincecum), December 3, 1860; Burkhalter, *Gideon Lincecum*, p. 138.

CHAPTER 7

1. John Henry Brown (1820–95) was a soldier in the Texas War for Independence and an Indian fighter on the frontier before becoming a prominent newspaper editor and legislator. A strong advocate of secession, he served as a Confederate officer in the Civil War and then promoted the immigration of ex-Confederates to Mexico. Later he earned fame as a Texas historian. "John Henry Brown," *New Handbook of Texas*, 1:765. Lincecum's diary for 1867 notes that on June 21, 1867, he was "reading John Henry Brown's discription of Tuxpan Valley, Mexico."

2. For a discussion of Gov. Throckmorton's dismissal, see Burkhalter, *Gideon Lincecum*, pp. 237–38 and nn.

3. In the nineteenth-century American West, the term "ambulance" was applied to a nonmedical "wagon with springs which gave a more comfortable ride." A. C. Greene, *900 Miles on the Butterfield Trail*, p. 86, n. 79.

4. One meaning of "yellow jack" is the yellow flag of quarantine, as applied to yellow fever. Decades later it was discovered that yellow fever is caused by a flavivirus carried by the mosquito, *Aedes aegypti*.

5. At this time Lincecum's sons Leonidas, Lucullus, and Lysander Rezin were practicing physicians in the Long Point area. Burkhalter, *Gideon Lincecum*, pp. 86–87.

6. Lincecum's diary for 1867 has the following entry for July 20: "Mr. Ruter, son of the founder of the college at Rutersville, Fayette County, Texas, called on me today and I had a very agreeable time with him. He is a highly educated gentleman, with a mind well stored with profitable knowledge."

7. Chocolate Bayou is a tributary of West Bay on the Gulf Coast in Brazoria County.

8. Burkhalter identifies Jimmy Fowler as being "a young man of Long Point." No Jimmy Fowler appears in the 1860 or 1870 census for Washington County. Burkhalter, *Gideon Lincecum*, p. 238.

9. "George" seems to be George Campbell, son of Leonora Lincecum Campbell and grandson of Gideon, who had previously accompanied Gideon on his expedition to West Texas in 1867. Ibid., pp. 231, 248.

10. Mr. Willbourn remains unidentified. Census records do not reveal him.

11. Gideon's "Christmas tune" was the Scottish ballad "Killiecrankie." See Lincecum and Phillips, *Adventures of a Frontier Naturalist*, pp. xxxi–xxxii.

12. Liverpool is a small town on Chocolate Bayou in Brazoria County, founded in 1837, and sparsely populated in 1868. "Liverpool, Texas," *New Handbook of Texas*, 4:247.

13. Apparently this Samuel Adams was a resident of Chappell Hill in Washington County. See U.S. Census, 1860, p. 220. Perhaps he was visiting friends or relatives in Brazoria County.

14. Lincecum's reference to "gen. Plateson" probably refers to a genus of the flounder. Flounders are in the family Pleuroneatidae. C. Richard Robins, *Common and Scientific Names of Fishes from the United States and Canada*, pp. 67, 105.

CHAPTER 8

1. Tuxpan, in full *Tuxpan de Rodriguez*, is a city in Veracruz state, east central Mexico, on the Rio Tuxpan, 7 1/2 miles from its mouth on the Gulf of Mexico. The precise location is 20° 57' N, 93° 07' W.

2. Bradford was a young doctor from Galveston who fell in love with Leonora Campbell's daughter Attilia (nicknamed Attie) and married her in Tuxpan on April 27, 1871. Burkhalter, *Gideon Lincecum*, pp. 273–74; Gideon Lincecum, "Diary," April 27, 1871, p. 40.

3. The chachalaca *(Ortalis vetula vetula)* is a large, brown, rather ugly bird that calls out its name raucously and monotonously every morning and before a rain. A few are found in the Texas Rio Grande Valley. Peterson, *Field Guide to Western Birds*, pp. 44–45.

4. Although young Willard R. Doran (1868–1921), Gideon's "Bully Grandson," was only two years old, Gideon addressed a number of his Tuxpan letters to him, having taken offense to a remark made by "Bully's" mother, Sarah, in one of her letters. In this way he could show his displeasure while still keeping touch with his dearest daughter. Burkhalter, *Gideon Lincecum*, pp. 256, 269–70.

5. The yellow butterflies are probably in the genus *Phoebis* Hübner, 1819. They are easily recognized by their general appearance, medium to large size, yellow or white coloration, and general abundance throughout the neotropics in all habitats. All of the species are migratory. Philip J. DeVries, *The Butterflies of Costa Rica and their Natural History*, pp. 102–104.

6. The "agua cate" is referring to the avocado, *Persea americana* Mill. Correll and Johnston, *Manual of the Vascular Plants of Texas*, p. 661.

7. *Anacampsis cerealella* has not been identified.

8. "The great blue winged beauty" may belong to the genus *Morpho* Fabricius, 1807. The brilliant coloration, found mostly in the males, ranges from pure white to azure to an intense, almost transparent blue violet. The males are more obvious because they patrol up and down rivers and forest edges. DeVries, *Butterflies of Costa Rica*, pp. 241–45.

9. The Portuguese Man-of-War *(Physalia)* is in the phylum Cnidaria (Coelenterata) and in the order Siphonophora. Barnes, *Invertebrate Zoology*, pp. 135–37.

10. The Banyan tree is *Ficus benghalensis* in the family Moraceae, which includes figs, hemp and mulberries. The uba (coba) tree may refer to *Coccoloba uvifera*, the sea grape, which is in the family Polygonales. "Uba" may be Lincecum's

spelling for *uva*, which means grape in Spanish. V. H. Heywood, *Flowering Plants of the World*, pp. 77–78, 96–97.

11. The "grampi" is Lincecum's designation of the genus *Grampus*, the dolphins.

12. Lincecum's term "coculus indicus" should be "cocculus indicus," which refers to a woody vine of the East Indies, *Anamirta cocculus* (family Menispermaceae). This plant produces berries which contain the poison picrotoxin. The berries will kill fish and can be used in hunting them. Heywood, *Flowering Plants of the World*, pp. 50–51.

13. The large white trumpet shaped convolvulus is probably a member of the Family Convolvulaceae which includes the morning glories, bindweeds, and sweet potato. Ibid., pp. 229–30.

14. "Roman camomile" refers to the plants of the asteraceous genus *Anthemis*, especially *A. nobilis* L. (the common camomile of Europe and of gardens elsewhere), an herb with strongly scented foliage and flowers which are used medicinally. *Gray's New Manual of Botany*, p. 846.

15. "Yerba dulce" refers to a sweet herb. In the Spanish language *yerba* is herb and *dulce* is sweet.

16. The term "umbrageous" means shade.

17. *Zapatito (Zapatilla) de la reina* is Spanish for "queen's little shoe." *The American Heritage Spanish Dictionary*. It is a yellow poppy. Lincecum is confused when he says it is a bean. The poppy belongs to the family Papaveraceae. *Argemone mexicana* L., the Chicalote or mexican poppy, is yellow; some authors say there is a second yellow poppy, *A. ochroleuca*. Charles T. Mason, Jr., and Patricia B. Mason, *A Handbook of Mexican Roadside Flora*, pp. 275–76.

18. The chacloco (coralillo) shrub is in the family Rubiaceae, the madder family. Its scientific name is *Hamelia patens* Jacq. Although often planted as an ornamental, it sometimes is considered weedy because it grows in disturbed areas along the roadside. Mason and Mason, *Handbook of Mexican Roadside Flora*, pp. 307–10.

19. The bonnet gourd is also known as "dishrag squash"; the fruit is about one foot long and closely netted in vascular bundles that resemble sponges in texture when skin, pulp, and seeds are removed. This gourd is in the family Cucurbitaceae, genus *Luffa*. Correll and Johnston, *Manual of the Vascular Plants of Texas*, pp. 1506–14.

20. For a treatment of Texas northers in Mexico, where they were called "El Norte," see Phillips, "The Texas Norther," Chap. III, "'El Norte' and the Mexican War," pp. 45–70.

21. Lysander Rezin Lincecum (1836–?) was Gideon's youngest surviving son, a practicing physician, who with his wife, Mollie, occupied Gideon's old home at Long Point. Burkhalter, *Gideon Lincecum*, pp. 230, 292, 295.

22. Ibid., pp. 292–93. Letter to W. P. Doran, July 8, 1873.

23. Lincecum's articles in *The American Sportsman*, entitled "Personal Reminiscences of an Octogenerian," ran from September 12, 1874, to January 16, 1875. An abridged version appears in Lincecum and Phillips, *Adventures of a Frontier Naturalist*.

24. Othniel Marsh (1831–99), one of the nation's foremost paleontologists, taught for over thirty years at Yale and led many fossil-hunting expeditions in the American West. George P. Merrill, "Othniel Marsh," *Dictionary of American Biography*, Vol. 6, Pt. 2, 302–303.

25. In addition to the citations by Coues, Joel Asaph Allen, several of whose monographs appeared in the same Hayden Survey volume, also referred to several of Lincecum's contributions. He cited notes on the water hare or muskrat, *Lepus aquaticus*, on pp. 281 and 335–36, and said (p. 366): "Dr. Lincecum calls it 'a widely distributed species,' and says 'it abounds on the canebrakes of Alabama, Mississippi, Louisiana, Arkansas, and Texas,' where it is found 'on all the water-courses, even on the little branches,' but 'rarely on the uplands.'" Coues's *Fur-Bearing Animals: A Monograph of North American Mustelidae* was published as No. 8 of the Hayden Surveys in 1877, with no contributions from Lincecum.

26. Dr. Andrew Weir had been a friend of Lincecum's in Columbus, Mississippi, and had emigrated to East Texas. He is listed as living in Titus County (Northeast Texas) in the U.S. Census for 1870, p. 002.

APPENDIX 4

1. Lampasas Sulphur Springs are located in the town of Lampasas, Lampasas County. "Lampasas County," *New Handbook of Texas*, 4:49–51.

2. The springs in Gonzales County led to the establishment there of Palmetto State Park and the Warm Springs Foundation for Crippled Children in the twentieth century. "Warm Springs Foundation for Crippled Children," *Handbook of Texas*, 2:863; "The Warm Springs Rehabilitation Center," *New Handbook of Texas*, 6:825; "Palmetto State Park," Ibid., 5:25.

3. Sour Lake in Hardin County, east of Houston, developed as a health resort in 1850 because of its medicinal waters, but declined as a resort when a major oil discovery was made in the area in 1902. "Sour Lake, Texas" *New Handbook of Texas*, 5:1150–51.

4. Piedmont Springs in Grimes County was a popular resort and health spa as early as 1850. Post–Civil War decline caused the spa to close in the 1870s. "Piedmont Springs," *New Handbook of Texas*, 5:192–93.

Selected Bibliography

RECORDS AND DOCUMENTS

Eighth Census of the United States (1860). Texas.

Lincecum Papers. Center for American History, University of Texas at Austin, Austin, Texas.

"1976 Annual Report of the Chief of Engineers on Civil Works Activities." Department of the Army, Office of the Chief of Engineers. 2 vols.

Ninth Census of the United States (1870). Texas.

Seventh Census of the United States (1850). Mississippi: Lowndes County.

BOOKS

Ajilvsgi, Geyata. *Wildflowers of Texas.* Bryan, Tex.: Shearer, 1984.

The American Heritage Spanish Dictionary. Boston: Houghton Mifflin, 1986.

Arnett, Ross H. *American Insects.* New York: Van Nostrand Reinhold, 1985.

Arnett, Ross H. *The Beetles of the United States.* Washington, D.C.: Catholic University of America Press, 1960.

Barnes, Robert B. *Invertebrate Zoology.* 4th ed. Philadelphia: Saunders, 1980.

Borror, Donald J., and Dwight M. DeLong. *An Introduction to the Study of Insects.* New York: Holt, Rinehart and Winston, 1954.

Brown, Clair A. *Louisiana Trees and Shrubs.* Baton Rouge: Louisiana Forestry Commission, 1945.

Burkhalter, Lois Wood. *Gideon Lincecum, 1793–1874: A Biography.* Austin: University of Texas Press, 1965.

Burkhardt, Frederick, and Sydney Smith, eds. *A Calendar of the Correspondence of Charles Darwin, 1821–1882.* New York: Garland, 1985.

Correll, Donovan S., and Marshall C. Johnston. *Manual of Vascular Plants of Texas.* Renner, Tex.: Texas Research Foundation, 1970.

Coues, Elliot, and Joel Asaph Allen. *Monograph of North American Rodentia.* F. V. Hayden, ed., *Report of the United States Geological Survey of the Territories.* Vol. 11. Washington: Govt. Printing Office, 1877.

Darwin, Charles. *The Origin of Species and The Descent of Man.* New York: Modern Library, n.d.

Davis, William B. *The Mammals of Texas.* Bulletin No. 41. Austin: Texas Game & Fish Commission, 1960.

DeVries, Philip J. *The Butterflies of Costa Rica and their Natural History.* Princeton: Princeton University Press, 1987.

The Dictionary of American Biography. Ed. Allen Johnson and Dumas Malone. 10 vols. and supplements. New York: Charles Scribners, 1958.

Elias, Thomas. *Complete Trees of North America.* New York: Van Nostrand, Reinhold, 1980.

Encyclopedia Americana. New York: Americana Corporation, 1958.

Field Guide to Birds of North America. Washington, D.C.: National Geographic Society, 1983.

Finsley, Charles. *A Field Guide to Fossils of Texas.* Austin: Texas Monthly Press, 1989.

Fleming, James Rodger. *Meteorology in America, 1800–1870.* Baltimore: Johns Hopkins University Press, 1990.

Geiser, Samuel Wood. *Naturalists of the Frontier.* Dallas: Southern Methodist University Press, 1937.

Gertsch, Willis J. *American Spiders.* New York: D. Van Nostrand, 1949.

Gleason, Henry A., and A. Cronquist. *Manual of Vascular Plants of Northeastern United States,* 2nd. ed. New York: New York Botanical Garden, 1991.

Gould, Frank W. *Common Texas Grasses.* College Station: Texas A&M University Press, 1978.

Gray's New Manual of Botany: A Handbook of the Flowering Plants and Ferns. Benjamin L. Robinson and Merrit L. Fernald, revs. 7th ed. New York: American Book Co., 1908.

Great Plains Flora Association. *Flora of the Great Plains.* Lawrence, Kans.: University of Kansas Press, 1986.

Greene, A. C. *900 Miles on the Butterfield Trail.* Denton: University of North Texas Press, 1994.

The Handbook of Texas. Ed. Walter P. Webb and Eldon S. Branda. 3 vols. Austin: Texas State Historical Association, 1952, 1976.

Harrar, Ellwood S., and J. George Harrar. *Guide to Southern Trees.* 2nd ed. New York: Dover, 1962.

Heywood, V. H., ed. *Flowering Plants of the World.* New York: Mayflower Books, 1978.

Hickman, Cleveland P., Sr., Cleveland P. Hickman, Jr., and Frances M. Hickman. *Integrated Principles of Zoology.* 5th ed. St. Louis: C. V. Mosby, 1974.

Hölldobler, Bert, and Edward O. Wilson. *The Ants.* Cambridge, Mass.: Harvard University Press, 1990.

King-Hele, Desmond. *The Essential Writings of Erasmus Darwin.* London: MacGibbon & Kee, 1968.

Lincecum, Jerry Bryan, and Edward Hake Phillips, eds. *Adventures of a Frontier Naturalist: The Life and Times of Dr. Gideon Lincecum.* College Station: Texas A&M University Press, 1994.

Mason, Charles T., Jr., and Patricia B. Mason. *A Handbook of Mexican Roadside Flora.* Tucson: University of Arizona Press, 1987.

McCook, Henry Christopher. *The Natural History of the Agricultural Ant of Texas*. Philadelphia: J. B. Lippincott, 1879.

The New Handbook of Texas. Ed. Ron Tyler, Douglas E. Barnett, and Roy R. Barkley. 6 vols. Austin: Texas State Historical Association, 1996.

Nowak, Ronald M. *Walker's Mammals of the World*. 5th ed. Vol. 1. Baltimore: Johns Hopkins University Press, 1991.

Peterson, Roger T. *A Field Guide to Western Birds*. 3rd ed. Boston: Houghton Mifflin, 1990.

Robbins, Chandler S., Bertel Bruun, and Herbert S. Zim. *A Guide to Field Identification of Birds of North America*. New York: Golden Press, 1966.

Robins, C. Richard, Chairman. *Common and Scientific Names of Fishes from the United States and Canada*. 5th ed. Bethesda, Md: American Fisheries Society, 1991.

Sorensen, W. Conner. *Brethren of the Net: American Entomology, 1840–1880*. Tuscaloosa: University of Alabama Press, 1995.

Stephen, Leslie, and Sidney Lee, eds. *The Dictionary of National Biography*. 21 vols. London: Oxford University Press, 1960.

Swann, Lester A., and Charles S. Papp. *The Common Insects of North America*. New York: Harper and Row, 1972.

Thomas, Clayton L., ed. *Taber's Cyclopedic Medical Dictionary*. 16th ed. Philadelphia: F. A. Davis, 1989.

Tyler, George W. *The History of Bell County*. Waco: Texian Press, 1966.

Wheeler, William Morton. *Ants: Their Structure, Development and Behavior*. New York: Columbia University Press, 1910.

Williams, F. L. *Matthew Fontaine Maury, Scientist of the Sea*. New Brunswick: Rutgers University Press, 1963.

Willis, J. C. *A Dictionary of the Flowering Plants and Ferns*. 8th ed. Rev. by H. K. Airy Shaw. London: Cambridge University Press, 1973.

ARTICLES, PERIODICALS, AND MANUSCRIPTS

Bradford, A. L., and T. N. Campbell. "Journal of Lincecum's Travels in Texas, 1835." *Southwestern Historical Quarterly* 53, no. 2 (1949): 180–201.

Cooke, John. "A Pioneering Spider Man." *Natural History* 105, no. 7 (1996): 74–75.

Heyman, I. Michael. "Smithsonian Perspectives." *Smithsonian* 27, no. 2 (1996): 22.

Lincecum, Jerry Bryan, and Edward Hake Phillips. "Civil War Letters of Dr. Gideon Lincecum: 'I am out and out a secessionist.'" *Texas Studies Annual* 2 (1995): 144–66.

McCook, H. C. "The Agricultural Ants of Texas." *Proceedings of the Academy of Natural Sciences* 11 (1877): 299–304.

Phillips, Edward H. "The Texas Norther." *The Rice University Pamphlet* 41, no. 4 (1955).

Sinks, Julia Lee. "Editors and Newspapers of Fayette County." *Quarterly of the Texas State Historical Association* 1 (July 1897): 37.

Wolfe, Cheri L. "The Traditional History of the *Chahta* People: An Analysis of Gideon Lincecum's Manuscript." Doctoral dissertation. University of Texas at Austin, 1993.

Index

Academy of Natural Sciences, 10, 13–14, 16, 28–29, 32, 33, 35, 48–49, 61, 92, 107

Adams, Samuel, 136, 198n. 13

Agrostis (winter bent grass), 59, 165, 187n. 9

Aix sponsa (wood duck), 101, 193n. 13

Algarobia glandulosa (mesquite), 68, 187–88n. 17

Alikchi chito, 51

Allen, Joel Asaph, 16, 200n. 25

alligator gar, 107–108

alum, 66–68, 188n. 6

American Entomological Society, 49, 101

American Naturalist, 14–15, 63–64, 80, 104, 106, 150

American Sportsman, 149–50, 199n. 23

Anacampsis cerealella, 141

Anser albifrons (white-fronted goose), 95, 192n. 3

ant battles, 10, 31, 35–37, 44–45

Anthemis (camomile), 145, 199n. 14

ant lion, 93–94, 191n. 42

ant rice. *See Aristida*

ants: agricultural (harvesting, stinging, *Pogonomyrmex*), 9–10, 19–23, 31, 37, 48, 50, 183n. 1, 185n. 19; —, battles of, 31, 44–45; —, cities of, 19–23, 31, 37–48; —, and founding of colony, 37–48, 50; —, mating of, 45–48; —, and prisoners, 40–41; —, stings of, 43; —, and workers, 38, 42, 47; "Bravo," 147–48; cutting (horticultural, *Atta*), 24–26, 31, 94, 183n. 1; —, battles of, 31; —, cities of, 24–26, 31; Gideon's catalog and, 184–85n. 17; large black tree (#1), 33–34; —, battles of, 35; microscopic red (#8), 32; red headed tree (#4), 36; —, battles of, 36–37; small black erratic (#5), 35–36, 49; —, battles of, 35–37; —, cities of, 36; —, and water, 24–26, 32–33, 41, 43–44

Ants of Switzerland, 48, 185n. 20

Ants: Their Structure, Development, and Behavior, 50

aptera, 191n. 43

Argemone mexicana (Mexican poppy), 26, 183n. 4

Aristida (ant rice), 21, 24, 39–40, 49–50, 183n. 2

Austin, Tex., 48, 61, 72, 117, 170, 195n. 5

Austin grass, 61, 187n. 11

Baird, Spencer Fullerton, 13–14, 16, 73–74, 93, 96–98, 101–104, 108, 116, 132, 152, 193n. 16

bark for dyeing cloth, 66
Barton, Benjamin Smith, 52, 186n. 3
Barton, William ("Uncle Billy"), 195n. 5
Barton's Creek (Tex.), 113, 195n. 5
Bastrop County, Tex., 192n. 7
bee moth (Galleria), 189n. 11
bees, 3, 5, 76–79, 188n. 9, 189n. 11
beetle. See Coleoptera
Bell County, Tex., 73, 192n. 7
Belton, Tex., 117
Bexar County, Tex., 177, 192n. 7
birds, 95–101
Birds of North America, The, 193n. 16
black bass, 108
Blanco County, Tex., 192n. 7
bonnet gourd, 147, 199n. 19
Botanical Textbook, 14
Botanic Garden, The, 9
Botany of the Southern States, 55
Bradford, Attilia ("Attie") Campbell, 141, 198n. 2
Bradford, George, 140, 198n. 2
Brazoria County, Tex., 197nn. 7, 12
Brazos River (Tex.), 170
British Honduras, 131
Bromus (Brome grass): B. ciliatus, 60, 187nn. 8, 10; B. corinatus, 60, 187nn. 8, 10
Brown, John Henry, 127, 193n. 19, 197n. 1
Brushy Creek (Tex.), 170
buckeye, Texas (Mexican), 67,187n. 17
Buckley, Samuel Botsford, 13, 28–29, 48–49, 62, 81–82, 95, 116, 128, 144, 149, 152–53
buffalo fish, 108
Burkhalter, Lois Wood, 7, 13, 181n. 3
Burleson County, Tex., 192n. 7
Burnet, Tex.: city of, 117, 172; county of, 73, 177, 192n. 7, 195n. 2
Busk, George, 26–27
Buteo regalis (ferruginous hawk), 97, 99, 192–93n. 11
butterflies (Lepidoptera), 91–92, 141–42, 198n. 5

cabbage bug, 90–91, 191n. 37

Calcarius (longspurs), 98, 192n. 9
Caldwell County, Tex., 192n. 7
Calendar of the Correspondence of Charles Darwin, A, 26
Cameron, Tex., 117
Campbell, George, 135, 137, 197n. 9
Campbell, Capt. George W., 182n. 17
Campbell, John, 143
Campbell, Leonora Lincecum, 138, 140, 182n. 17, 197n. 9
Capella gallinago (common snipe), 101, 193n. 13
Carya olivaeformis [illinoensis] (pecan), 67, 187n. 17
catfish, 108
Celtis occidentalis (hackberry), 26, 183n. 3
centipede, 84–86, 189n. 24, 190nn. 26–29
Cerasus cardinaensis (caroliniana?) [renamed Prunus caroliniana] (wild peach or laurel cherry), 68, 187–88n. 17
Chachalaca (Ortalis vetula vetula), 141
Chocolate Bayou, Tex., 133, 197nn. 7, 12
cicada, 72, 188n. 3
climate cycles, 123–24
Coccoloba uvifera (sea grape), 143, 198–99n. 10
"coculus indicus," 144, 199n. 12
Coleoptera, 73, 89, 116, 188n. 5, 190nn. 34, 35
Colorado River (Tex.), 61–62, 110, 113, 171, 177, 195nn. 2, 5
Columbus, Miss., 115
Comal County, Tex., 176, 192n. 7
Convolvulaceae, 144–45, 199n. 13
Cope, Edward Drinker, 14, 16–17, 107, 133, 153–54, 194n. 25
Coryell County, Tex., 73, 170, 192n. 7
cotton, 57
Coues, Elliot, 16, 149–50, 154
cougar, 103
Cow House Creek (Tex.), 117, 170
cow killer, 93–94, 191nn. 41, 42
Cox, Solomon B., 102–103, 193n. 19
Cresson, Ezra Townsend, 14, 49, 80, 89–90, 101, 154, 190–91n. 36

Cross Timbers, 112, 187n. 15
Cygnus columbianus (tundra swan), 95, 192n. 3
Cynodon dactylon (Bermuda grass), 59, 187n. 9

daddy-long-legs, 88–89, 190n. 33
Darwin, Charles, 7–8, 18, 23, 26–28, 30, 48, 131, 154, 183n. 7
Darwin, Erasmus, 8–9, 18–19, 30, 64
Daubentia (Spanish buckeye), 67
Descent of Man, The, 28, 183n. 7
Dickinson's Bayou, Tex., 137
dirt (mud) dauber (Sphecidae), 81–83, 189n. 16
Doran, Sarah Matilda ("Sallie") Lincecum, 13, 29, 128–29, 137–38, 140–45, 147–49, 151, 154, 184nn. 8, 9
Doran, Willard Richardson, 198n. 4
Doran, William P. ("Sioux"), 51, 130, 138, 154–55
Douglas, John H., 108
drought of 1860, 115, 123–24, 126
Dunn, Dr. W. A., 59, 78
Durand, Èlie (Elias) M., 10, 13–14, 17, 28–33, 35, 37, 62, 76, 80, 103, 117–18, 125, 127, 129, 138, 155, 184n. 8
Durham, Cassandra L., 133, 155
Durham, George J., 71, 99–102, 123, 155

Eaton, Amos, 186n. 4
Elanoides fortficatus (swallow-tailed kite), 97, 192n. 6
Englemann, George, 14, 51, 62, 129, 155
"epizooty," 72, 188n. 4
Essex Institute (Salem, Mass.), 14, 63, 157

falcon, 65
Falls County, Tex., 192n. 7
Falls of the Brazos (Tex.), 97, 116, 118
Festuca (tall fescue), 187n. 8
Ficus benghalensis (Banyan Tree), 143, 198n. 10
fish. See specific types
Fisher's Falls (Tex.), 177

fish kill, 144, 199n. 12
Fitch, Asa, 191n. 44
Fitzroy, Robert, 124–25, 155
flint, 111
Flint, Charles L., 191n. 44
Flora of the Southern United States, 14, 55
flounder, 137
Forel, Auguste, 48–49, 185n. 20
Forshey, Caleb G., 115, 120–22, 124, 155–56
Fort Bend County, Tex., 133
Fort Inge (Tex.), 177
fossils, 109, 111, 113, 116, 194n. 28
Fowler, Jimmy, 133, 135, 197n. 8
Fredericksburg, Tex., 123

Galveston Weekly News, 52, 125
gammer grass, 165
garden plants, 90, 126
Geiser, Samuel Wood, 195n. 7
Geococcyx californianus (chapparal cock or roadrunner), 95–96, 192n. 4
Geological Survey of Texas, 113–14
Gillespie County, Tex., 192n. 7
Gonzalez County, Tex., 175, 192n. 7
grasses, 56–62, 163–67
grasshoppers, 71–76, 141, 188nn. 1, 7
Gray, Asa, 63–64, 156
Grimes County, Tex., 176, 200n. 4
Grus americana (whooping crane), 99, 193n. 12
Guadaloupe County, Tex., 192n. 7
Guadaloupe River (Tex.), 62, 177
gypsum, 169–71

Haller, Karl, 192n. 8
Hamelia patens (chacloco, coralillo), 147, 199n. 18
Hamilton's Creek (Tex.), 117, 172–73
Hannay, R. B., 52, 156
Hardin County, Tex., 200n. 3
Harris, Thaddeus William, 191n. 44
harsh vine grape, 62
Hawkins, Tex., 117
hay crop, 57–59, 61
Hayden Survey (1877), 16, 150

Hays County, Tex., 192n. 7
Henry, Joseph, 13–14, 97, 125, 156
Hentz, Nicholas M., 189n. 13
Hölldobler, Bert, 10, 50
horned frog (*Phrynosoma*), 107, 194n. 24
horse chestnut, 187–88n. 17
Houston, Sam, 116
Houston Telegraph, 35
How to Collect and Observe Insects, 14
Huber, François, 76, 188n. 10

Ichneumonidae, 188n. 4
Ilex vomitoria (yaupon), 26, 183n. 3
Indians, 117, 102–103. *See also specific tribes*
Introduction to the Study of Natural History, 14
iron mountain (Tex.), 115, 195n. 7

Jardin des Plantes, 28
Jerusalem oak, 75, 188n. 8
Johnson, T. E., 62
Journal of the Linnean Society, 27, 50
Junco hyemalis (dark-eyed junco), 99, 192–93n. 11

"Killiecrankie," 197n. 11
Kuechler, Jacob, 123, 196n. 2

La Bahia Prairie, 57
Lampasas (Tex.): city of, 174; county of, 177, 192n. 7, 200n. 1; sulphur springs of, 173, 177, 200n. 1
Lampasas River (Tex.), 117, 173, 177
Lane, A. G., 81, 118, 189n. 17
Lanius ludovicianus (loggerhead shrike), 98, 192–93n. 11
Leidy, Joseph, 14, 16–17, 33, 35, 37, 73, 117, 156
Leon River (Tex.), 117
Lepidoptera (butterflies), 91–92, 141–42, 198n. 5
Libellulidae, 191n. 40
Lincecum, George W., 102, 192n. 10, 193n. 17
Lincecum, Gideon: and academic scientists, 9–10, 15–16, 33, 48–49, 55,

58, 61, 92, 144; as agronomist, 56, 58–59, 65–70, 90; and ants, 9–10, 18–50, 184–85n. 17, 186n. 22; and artesian well, 114–15, 197n. 6; attitude of, toward Charles Darwin, 24, 30, 131; attitude of, toward nature, 52, 55–56, 70, 92; attitude of, toward science, 11, 14–15, 17, 24, 27, 29–30, 37, 72, 93, 115, 140; attitude of, toward Spencer Baird, 101; and bee training, 76–77, 79; and biographical data, 7, 16; and botany, 14, 29, 51–70, 184n. 10, 186nn. 4, 6; and children, 128, 132–33, 138, 149; and Civil War, 15, 23, 26, 30, 32, 82, 116, 182n. 13, 197n. 6; and clairvoyance, 64–65, 121; on conservation, 56–57, 61–62; declining health of, 149–50; and doctor sons, 130, 197n. 5; on dyeing cloth, 66–68; early life of, 7–8, 24, 29, 49, 51, 193n. 18; and East Texas expedition (1867–68), 129, 132–33, 135–38; and entomology, 14–17, 71–79, 81–84, 88–94; and evolution, 18–19, 79, 109–10, 113, 118; and geology, 110–18; as graphologist, 28, 30; and Gulf expeditions, 99–100, 108; and imaginative thinking, 64–65, 82; and Indians, 102–103, 186n. 1; and indigenous plants, 52, 56–59, 61–62, 186n. 7; isolation of, from other scientists, 24, 29, 94; and "Lincecum Myth," 9–10, 21, 39–40, 49–50; lyrical writing of, 72–73, 92, 118, 120–22, 136–37, 151, 173; and mammals, 101–107, 149–50, 200n. 25; and medicine, 8, 32, 51–52, 62–63, 84–86, 130, 175; and meteorology, 4–5, 86, 95, 119–26; and music, 135; and natural resources of Texas, 114, 116, 176–78; and Northern scientists, 13, 29–30, 37, 61; personal habits of, 3, 136; personality of, 3, 6, 15, 130, 132, 138, 144, 149, 173; published writings of, 179–80,

194n. 21; reading habits of, 8, 14, 18, 51, 55, 99, 124; and Reconstruction, 56, 126–29, 131; and religion, 22–24, 35, 55, 82, 135, 151; and science methodology, 9, 15, 21, 41, 45–46, 49, 63–64, 66–68, 74–75, 89, 91–92, 112; scientific correspondents of, 11, 13–14, 152–58; and surf bathing, 143–44; and "water cure," 175; and West Texas expedition (1867), 97, 103, 116–17, 192n. 7; and wine making, 68–70; writing style of, 27, 49, 65

Lincecum, John P., 126, 156

Lincecum, Leonidas, 45, 130, 197n. 5

Lincecum, Lucullus, 130, 197n. 5

Lincecum, Lysander Rezin ("San"), 101, 130, 149, 193n. 13, 197n. 5, 199n. 21

"Lincecum Myth," 9–10, 21, 39–40, 49–50

Lincecum Papers, 16, 181nn. 1, 3, 186n. 4, 192n. 2

Lincecum, Sarah Bryan, 3, 108–109, 116, 118, 127–28, 194n. 26

Lincecum, Sarah Matilda ("Sallie"). See Doran, Sarah Matilida ("Sallie") Lincecum

Lindheimer, Ferdinand, 14

Linnean Society, 26

Liverpool, Tex., 136, 197n. 12

Llano County, Tex., 115, 177, 195n. 7

Llano Estacado, 170

lobelia, 130

Long Point, Tex., 3, 21, 64, 95, 97, 110–12, 115–16, 120–26, 195n. 4

marble, 112

Marble Falls, Tex., 111, 171, 177, 195n. 2

Marcy, Captain Randolph, 171

Marsh, Othniel, 149, 200n. 24

Matson, Mary Lincecum, 117, 156

Maury, Matthew F., 122, 124, 156

McCook, Henry C., 48–50, 156–57, 183n. 2

medicated waters of Texas, 173–76

mesquite grass, 165, 166

Mexican plants, 141, 144–47

Milam County, Tex., 192n. 7

Mitchell, I. A., 59

Moggridge, John Traherne, 48, 185n. 19

Moore, Daniel Boone, 138, 157

Moore, Emily Lincecum, 138, 157

Moore, Francis, Jr., 115, 195n. 7

Mormon Mills, Tex., 117, 196n. 12

Morpho, 198n. 8

Morus (mulberry), 186n. 6

Mount Bonnell (Tex.), 170

Mount Olympus (Gideon's homeplace), 123, 126

Muridae, 16, 150

Muscogee, 51

Mygale Hentzii (tarantula), 80–81, 189n. 13

myriapoda, 15

Myriapoda of North America, 14

National Observatory, 125

Natural History of the Agricultural Ant of Texas, 48

natural selection, 18–19, 24, 28, 30, 131

negus tree, 83, 189n. 22

New Braunfels, Tex., 176

Noland's Creek, Tex., 117

norther, 73, 75, 119–23, 126, 138, 147–48, 196n. 5, 199n. 20

Numenius (curlew), 101, 193n. 14

oats, wild, 163

On the Origin of Species, 18, 28, 131

opossum, 104–106

Ortalis vetula vetula (chachalaca), 141, 198n. 3

owl, 98, 192n. 11

Owl Creek (Tex.), 117

ozone paper, 121, 124, 196n. 4

Packard, Alpheus S., Jr., 14, 17, 63, 128, 157

Panicum gibbum (*P. obtusum*; Austin grass), 61, 187n. 11

paper wasp (Vespinae), 83–84, 189n. 21

Parker, H. C., 131

Passerherbulus caudacutus (LeConte's sparrow), 98, 192n. 9

Peck, George W., 17, 91, 157
perch, 108
petrified wood, 112
Phalaris intermedia (canary grass), 60, 187n. 10
Phoebis, 198n. 5
Pholadomya lincecumi, 194n. 28
Physalia (Portuguese Man-of-War), 142–43, 198n. 9
Physical Geography of the Sea, 124
Piedmont Springs (Tex.), 176, 200n. 4
pigeon, wild, 64–65
plants, poisonous, 52–55
Plectrophaenax nivalis (snow bunting), 99, 192–93n. 11
Pogonomyrmex molefaciens. See ants: agricultural
Polioptila (gnat catcher), 99, 192–93n. 11
Pooecetes gramineus (vesper sparrow), 98, 192n. 9
Port Sullivan, Tex., 117
Practical Entomologist, 14, 90, 190–91n. 36

Quercus (oak): practical uses of, 65; *Q. catesby (Q. laevis)* (turkey) 97, 192n. 5; *Q. obtusiloba* (post), 63–64, 123, 187n. 15; *Q. palustris* (pin) 124, 196n. 3; *Q. Tinctoria* (black), 64–65, 67, 187n. 16

rat, 106–107
Record of American Entomology, 17
redbugs, 15
red fish, 137
Redshaw, Peggy A., 11
Report on the Insects of Massachusetts Injurious to Vegetation, 191n. 44
rescue grass, 57, 164–65
Richmond, Tex., 133
Round Rock, Tex., 117, 170
Ruff, D. E., 98, 192n. 10
Ruter, Mr., 131, 197n. 6
Rutherford, M., 125
rye grass, 164, 167

salamander *(Geomys bursarius)*, 107, 194n. 23
San Antonio Springs (Tex.), 177
San Carlos, 138
San Gabriel River (Tex.), 117
San Marcos Springs (Tex.), 177
"scarabacus," 189n. 20
science: in nineteenth-century America, 10–11; in Texas, 13, 29
scorpion, 86–88, 190n. 32
sea shells, 108–109, 116
shrews, 103–104, 193–94n. 20
Shropshire, Ben, 61, 157
Shumard, Benjamin F., 116, 157, 195n. 8
Smithsonian Institution, 13–16, 73, 96, 98, 100–102, 108, 116–17, 125, 133, 150, 182n. 15, 191n. 39, 193n. 17
Somerville Lake (Tex.), 195n. 4
Sorensen, W. Conner, 16–17, 28
Sour Lake (Tex.), 175, 200n. 3
Southern Cultivator, 55, 58, 61, 77
Spanish moss, 63–64, 68, 88
specimens, preservation and shipping of, 73, 92, 98, 132–33, 182n. 15, 184n. 10, 191n. 39
Sphyrapicus varius (yellow-bellied sapsucker), 99, 192–93n. 11
spiders, 80–81, 86–87, 120, 189nn. 13, 14, 15, 190n. 30. *See also specific types*
spider wasp (Pompilidae), 81, 189nn. 18, 19
Spillman, William, 109–10, 194n. 29
Stafford's Point, Tex., 133
State Geological Survey, 116
Stewardson, Thomas, 30
Stipa setigera (mesquite or green wintergrass), 59–60, 187n. 9
Strachia histrionica (cabbage bug), 191n. 37
Summers, S. H., 102
syrup making, 68–70

Talpa (moles, shrews), 107, 194n. 23
tarantula, 80–82
Taylor, Mary H., 66
Temple of Nature, The, 9, 18
Tephrosia virginiana, 85, 190n. 25

Texas Almanac, 52, 56, 61, 92, 116
Throckmorton, James Webb, 116, 128, 158, 195n. 11, 197n. 2
Tidwell, David, 190n. 31
Tillandsia usneoides (Spanish moss), 63–64, 69, 88
Tombecbee River. *See* Tombigbee River
Tombigbee (Tombecbee) River (Miss.), 111, 195n. 3
Travis County, Tex., 192n. 7
Trifolium Texana [repens] (white clover), 60, 187n. 10
Tuxpan, Mexico, 101, 127, 131–32, 138–49, 198n. 1

Ulmus americana (broad-leaf elm), 68, 187–88n. 17
Unios, 172
Uvalde County, Tex., 177

Vaughan, George H., 68
Veracruz State, 127
Viburnum dentatum (arrowweed or prairie dogwood), 26, 183n. 3
Vitis (grape). *See specific types*
V. lincecumii (post oak), 62, 187n. 13
V. mustangensis (mustang), 26, 62, 69, 187n. 13, 183n. 3;

Walsh, Benjamin D., 190–91n. 36

Washington County, Tex., 3, 192n. 7, 195n. 4
wasp. *See specific types*
Weems, Mason Locke, 8, 182n. 5
Weir, Andrew, 151, 200n. 26
West Bay (Tex.), 197n. 7
wheat, 74–75, 164
Wheeler, William M., 50, 183n. 2, 186n. 23
Willbourn, Mr., 135
Williamson County, Tex., 170, 192n. 7
Wilson County, Tex., 177
Wilson, Edward O., 10, 50
Wilson, W. A., 55, 65, 70
wine making, 68–70
Wixson Creek (Tex.), 175
Wolfe, Cheri Lynn, 186n. 1
Wood, Horatio Charles, Jr., 14–15, 37, 80, 83, 85–89, 92–93, 107, 158

Xanthoxylum (Zanthoxylum) carolinianum (prickly ash), 26, 183n. 3

Yegua Creek (Tex.), 111, 195n. 4
yellow fever, 93, 129
yellow jacket (Vespinae), 83, 189n. 23

Zoonomia, 8